MARGINS OF REALITY

The Role of Consciousness in the Physical World

Robert G. Jahn *and* Brenda J. Dunne

A Harvest Book

HARCOURT BRACE & COMPANY

San Diego New York

**To Toby and to Sam;
To Gretchen, Nimbus, Macho, and Shadow;
To the Deer and the Geese;
And to all our other animal friends
Who kept watch, and understood it all**

Library of Congress Cataloging-in-Publication Data
Jahn, Robert G.
Margins of reality.
Bibliography: p.
1. Consciousness. 2. Parapsychology and science.
I. Dunne, Brenda J. II. Title.
B105.C477J34 1987 133.8 87-10041
ISBN 0-15-157148-1
ISBN 0-15-657246-X (pbk.)

Designed by G. B. D. Smith
Printed in the United States of America
G I J H F

MARGINS OF REALITY

CONTENTS

SECTION FIVE

The Vectors Deflected *289*

PREAMBLE

Consciousness contemplates no more profound or perplexing question than this: what is its role in the establishment of reality? In one extreme view, endorsed by much of traditional Western science and philosophy and exemplified by modern pragmatic materialism, the mind of man is relegated to a passive processor of experience imposed by a totally deterministic external world—a mere visitor meandering through the grand museum of life. In the opposite extreme, espoused by numerous and enduring mystical traditions of many cultures and eras, all experience is presumed to be created by the consciousness, so that any tangible reality ultimately traces to illusion. Between these divergent perspectives there is room for a full variety of hybrid personal positions, to be formed by some constellation of cultural and religious heritage, empirical experience, logical reasoning, intuition, and inspiration, wherein consciousness is allowed some mixture of these passive and active roles—wherein, as Niels Bohr concluded from the enigmas of modern physics, "we are both onlookers and actors in the great drama of existence."[1]

Until quite recently, scholarly search for the most satisfying and pragmatic balance of these perspectives has necessarily proceeded largely on hypothetical and qualitative grounds, for directly illuminating experimental evidence has been very difficult to obtain. Although alchemists of every age have expended immense efforts in attempting to break this cosmic riddle, their results have only compounded the question. While the modern disciplines of academic and clinical psychology have collected many empirical correlations and proposed numerous heuristic models of the interactions of human consciousness with its environment, the underlying mechanisms of the dialogue between mind and matter remain far from comprehended. On the physical side as well, few directly pertinent experimental data exist, although an assortment of anomalous effects such as the measurability paradoxes of atomic and nuclear physics, the enigmas of neuroscience and medical practice, and the body of controversial evidence on so-called psychic phenomena suggest that there may be serious shortfalls in the established paradigms.

Given the immense pragmatic and philosophical ramifications of the topic, this dearth of definitive experimental evidence might seem curious, but formidable obstacles oppose such investigations. The physical and psychological relationships between consciousness and its physical world entail subtle effects and processes that in some cases appear to violate the most fundamental scientific premises of space, time, and causality. The parameters bearing on such interactions are numerous and widely interdisciplinary, and various subjective and aesthetic factors not normally accommodated by traditional scientific methodology seem crucially relevant. Even in the most incisive and carefully controlled studies, systematically replicable effects appear only as marginal deviations from the normal statistical behavior of probabilistic systems, predicating extreme sensitivity and stability of detection equipment and acquisition of huge data bases if valid indications are to be unequivocally discriminated from the intrinsic noise of random processes.

With full acknowledgment of these difficulties, the purpose of this book is to reexamine the role of consciousness in the establishment of physical reality in the light of an extensive new body of experimental data recently acquired using equipment, protocols, data-processing and interpretation techniques that mitigate the obstacles to the extent that statistically replicable, quantitative results can be obtained. Some of these experiments study the interaction of human operators with various technical devices and systems. Others concern the acquisition of information about remote geographical targets inaccessible by known sensory channels. From these studies emerge a family of manifestly anomalous effects, clearly at variance with established physical and psychological models, and clearly indicative of some influence of human consciousness on the tangible statistical characteristics of the information involved. Based on the observed salient features of these anomalies, their scale and primary parameters, a conceptual model of such interactions can be proposed to aid in correlation of results, design of more incisive experiments, prediction of other effects, and explication of the phenomena on fundamental grounds. While the random processes embodied in these experiments and presumed in this model are specifically selected and adapted for rigorous and quantitative research, they share many basic characteristics with numerous other probabilistic systems in the natural world, in psychological and sociological situations, and in engineering practice. Hence, various broader implications of the observed and postulated effects can reasonably be considered.

The thesis is posed in five segments. In Section I are assembled a group of epistemological vectors, drawn from various traditional areas of human thought and experience, that converge on the motivation, definition, and circumscription of this particular research program. Section II presents a concise review of the design, operation, and results of an ensemble of experiments directly addressing the interaction of human consciousness with physical devices and systems embodying various types of random processes. Section III describes a complementary program of experiments on the acquisition and interpretation of information about unknown physical targets remote in space and time. In Section IV, these demonstrated anomalous capabilities of human consciousness are addressed via a comprehensive theoretical model based on the same wave mechanical metaphor that proved useful in resolving many of the anomalies of modern physics. Section V then reexamines the original set of topical vectors in the light of these experimental results and theoretical predictions and attempts to converge on some generic response to the original question of the role of consciousness in the establishment of reality.

Many technical and philosophical fibers are required to weave these arguments, some quite rudimentary and intuitive, others more sophisticated and hierarchical, and their interactions are equally as important as their individual courses. But once this conceptual array is in place and the empirical evidence is sifted through it, the role of consciousness in the physical world indeed emerges endowed with an active component. By virtue of the fundamental processes by which it exchanges information with its environment, orders that information, and interprets it, consciousness has the ability to bias probabilistic systems, and thereby to avail itself of certain margins of reality.

MARGINS
OF
REALITY

SECTION ONE

The Vectors

Sic quasi membranæ volitant Simulacra per auras.
Quaq; patet quocunq; licet conjuncta feruntur

Assuredly, all the disciplines of the mind and all the sciences of man are equally precious and their discoveries mutually so, but this solidarity does not mean confusion. What is important is to integrate the results of the diverse applications of the mind without confounding them. The surest method . . . is still that of studying a phenomenon in its own frame of reference, with freedom afterwards to integrate the results of this procedure in a wider perspective.

—MIRCEA ELIADE, *Myths, Dreams, and Mysteries*[1]

The Difficulty of Scientific Observation (18th Century)

1

SCIENTIFIC TWO-STEP

The bodies of established science move forward on two feet. One foot we call "experiment"—an observation or measurement performed under controlled conditions to acquire information about a natural effect or process. The other foot we call "theory"—a stated model, principle, or formalism to explain, correlate, or predict observational experience. When these two feet are sound, well balanced, and working in concert, and when the topical terrain they tread is propitious, science may move productively ahead, shifting its weight successively from empirical experiment to explicating theory, or from predictive theory to confirmatory experiment, in a brisk martial gait, a steady stroll, or a graceful swirling dance. But if either foot is lame, or is clad in technical or logical gear unsuited to the environment, or if the path extends through particularly obscure or hostile territory, progress is more tortuous and halting. Then each step must be tentative, first testing its ability to support the next advance of its counterpart and, when insecure, retreating to firmer position. In such hazardous conditions, experiments that yield unreliable or inadequate data can cause the theoretical foot to lurch off balance, or can inhibit its step altogether. Likewise, ill-posed theories that improperly acknowledge the empirical data, or that are inherently inadequate to deal with the phenomenological ground, may impel subsequent experimental steps into unproductive

Following the Footprints of Nature (17th Century)

directions. Avoidance of such digressions, or survival of them when unavoidably encountered, requires technical skill and experience, complemented by intuition, imagination, and inspiration.

Particularly critical in such excursions of science is its reaction to any anomalous features encountered by either of its feet: empirical observations grossly inconsistent with previous theoretical steps, or new theoretical predictions that conflict with previously established experimental data. Such anomalies stand like road signs, signaling "detour" or "wrong way" to the scientific traveler, and forcing him to ascertain which foot is on soggy ground. Is the anomaly an artifact of flawed experimentation, or is the theory inadequate to accommodate the full weight of the observed phenomena? Error in this discrimination will send the science along a false trail, from which it must eventually retreat if it is not to become mired. Proper assessment and assimilation of the anomaly, on the other hand, will point to a more penetrating path than that previously followed, along which the science can resume its progress more vigorously than before.

This two-step dynamic of science is, of course, just a particularly disciplined form of the more common process employed by human consciousness in establishing, ordering, and interpreting its personal reality. From birth to death, each individual accumulates experience, either incidentally or by design, and endeavors to interpret it and apply it to prediction of, or accommodation to, future events. Although the mechanics of acquisition, processing, and interpretation of the input data may be less sharply defined and structured than that of formal science, the alternating stride between experience and reason is just as essential to orderly progress. The role of anomalies encountered along this personal experiential trail is also just as critical. Events that arise to contradict established belief or habit pose a hierarchy of questions, first challenging the validity of those events, and then the prevailing pattern of personal convictions, until a more comprehensive resolution is achieved.

The burgeoning technical fields of artificial intelligence, expert systems, robotics, and automated manufacturing attempt to mimic much of this same strategy. After an empirical data base is established in the "memory" of a computational device or system, a sequence of programmed algorithms alternately deduce and apply "intelligent" exercises to accomplish particular tasks, while other routines attempt to improve the quality and quantity of the memory on the basis of the success of those actions. Like their human counterparts, much of the effectiveness and sophistication of such intelligent machines depends on their capacity to deal constructively with anomalies that arise in their own experience/logic loops that are not resolvable by routine search-and-compare exercises.

One can identify similar experience/logic dynamics on much larger collective and societal scales, as well. For example, the historical development of cultural strategies, principles, and values can be viewed as an alternation of relatively passive periods of accumulating events, followed by eras of structural

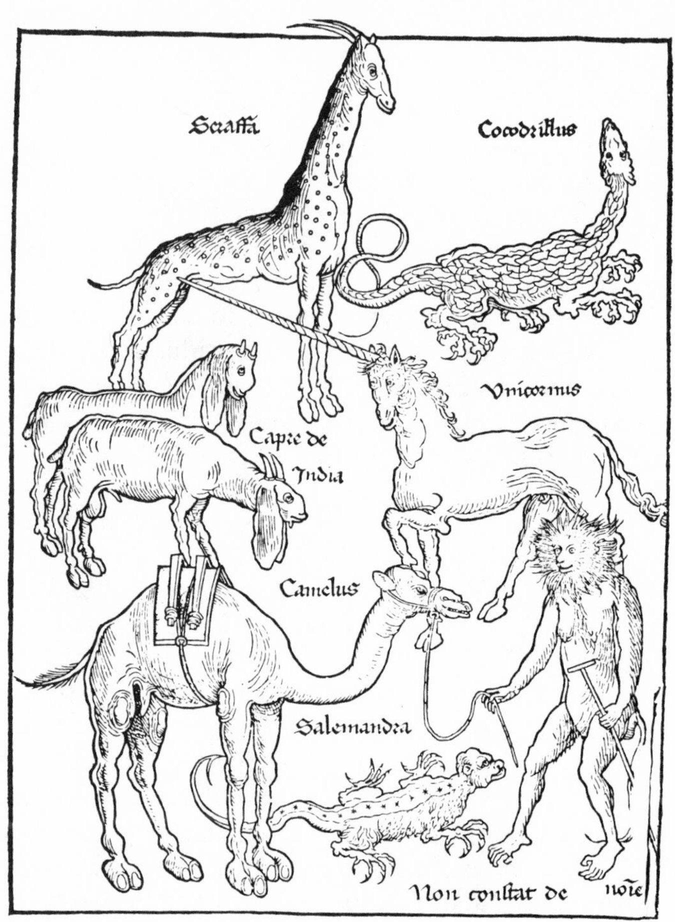

Hec animalia sunt veraciter depicta sicut vidimus in terra sancta

Anomalous Animals (15th Century)

and strategic changes responsive to those experiences. Here again, catastrophic major anomalies, such as wars, plagues, economic collapses, political or religious upheavals, indicative of basic flaws in the prevailing social situations, have precipitated the most severe redirections of the grand historical journey. The course of biological evolution could be similarly represented as a succession of genetic responses of organisms to challenges or changes in their environment.

Most anomalous phenomena that confront human logic, whether in the scientific, personal, or public domains, tend to have only local impact within their particular provincial contexts, and to arise and be resolved relatively quickly compared to the overall evolutionary paces of those provinces. In rarer instances, such as the anomalous celestial observations that contradicted the prevailing geocentric models and associated religious dogma, or the array of atomic-scale physical data that precipitated the quantum revolution, the implications can stretch more broadly across all three domains, and the efforts toward resolution then become more widespread, protracted, and intense.

The one genre of anomalous human experience that has dwarfed all others in its ubiquity, endurance, implications, and resistance to rational comprehension is that associated with the very processes of consciousness upon which all observational and deductive skills are based. Throughout recorded history, anecdotal instances of inexplicable consciousness-related phenomena have regularly been reported and variously interpreted in such diverse contexts as religion, philosophy, psychology, anthropology, medicine, art, and even high technology. Such events have been catalogued as "miracles," "magic," "psychic phenomena," "sorcery," "inspiration," "gremlins," and countless other elusive concepts, but little coherence has ever been

"Hence I have yielded to magic to see . . . what secret force hides in the world and rules its course"

Goethe: Faust
(17th-Century Woodcut)

established among them. Millennia of scholarly attention to such topics have failed to provide experimental data bases of adequate size and rigor to define this body of anomalous phenomena with any precision, let alone to support a scientific dialogue to resolve it. Nonetheless, in very recent years, a new generation of sensitive and efficient experimental tools has become available that permits more credible demonstration, specification, and correlation of such effects under controlled research conditions. As these more reliable empirical data compound, it now becomes feasible to propose complementary theoretical steps, and thereby to move cautiously toward some incorporation of these anomalies within a better general comprehension of the interaction of consciousness with its physical world.

In this book we shall describe the strategy, implementation, results, and interpretation of one such contemporary scientific safari into a portion of the phenomenological outback populated by a bewildering array of anomalies associated with the human operation of sensitive information-processing equipment and techniques. While the original orientation of this particular study was primarily technological, it rapidly became clear that the design of effective experimental methods to display and correlate such effects, the theoretical representation and interpretation of the data, and the broader scientific, personal, and social implications of the experimental and theoretical findings all required inclusion of many less tangible factors not usually admitted into the chambers of modern science. In fact, one of the most exciting aspects of this study has been its evident relevance to, and dependence upon, virtually all other lines of scholarly exploration.

In this sense, we might say that a number of diverse topical vectors have served to triangulate this research program—some historical, some contemporary; some philosophical, some technological; some intuitive, some intellectual. For example, the generic role of anomalies in the progress of science that we have just been discussing already qualifies as one such vector. Over the remainder of Section I we shall trace a few of the most influential other vectors from their conceptual sources to their intersection on our program. Then, after presenting the body of our experimental and theoretical work, we shall return, in the final section, to assess the possible effects of this research on those same vectors and attempt to weave them into a coherent conceptual container. While no one of them, in and of itself, will be fully persuasive in predicating or explicating this research, their complementary confluence will provide a serviceable framework for contemplation of an issue of the highest ontological and functional significance: nothing less than the role of human consciousness in the mechanics of reality.

2

MAN AND THE MYSTICAL

The most pervasive, yet most enigmatic vector we must acknowledge comprises nothing more, nor less, than the archetypal propensity of *Homo sapiens* to participation in a wide assortment of activities we shall subsume under the term "mystical." Within this category we include all manner of ceremony and ritual, oral and literary invocation, architectural and artistic effort, or any other form of spiritual commitment that is responsive to some hope, suspicion, or conviction of supernatural influences, or is deployed in some attempt to affect the course of physical events, to acquire information or insight, or to achieve transcendent spiritual experience. In short, this vector traces the role of abstract human desire in the establishment or perception of reality, which itself constitutes a generic anomaly in any deterministic paradigm.

By this definition, the mystical vector traces far back into the haze of prehistory. The ceremonial burial sites of the Neanderthals; the Paleolithic cave drawings of southwestern Europe; the Mother Goddess figurines spanning countless cultures and tens of millennia; the megalithic stone circles, monuments, henges, and barrows; the pyramids of Egypt, India, South and Central America; the oracle bones of the early Orient; the masks, petroglyphs, and totem poles of native American, African, and other aboriginal tribes—all testify, by their vastness, ubiquity, or social centrality, to an immense importance of the mystical dimension in otherwise fiercely

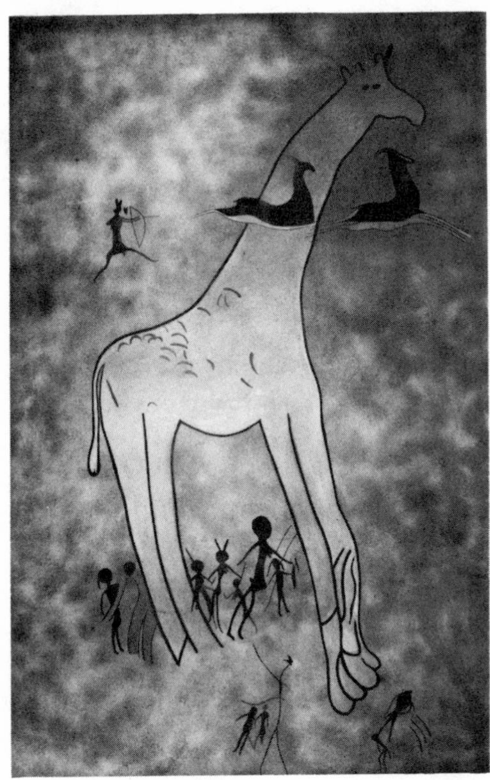

Tassili Wall Painting from the Libyan Sahara: The Great Giraffe Adjefou (ca. 3500 B.C.)

pragmatic primitive lives. The earliest creation stories, myths, heroic epics, fairy tales, folklore, and divinely inspired legends that survived from oral traditions into classic literature and art featured mystical allusions, superhuman characters, and miraculous processes that bridged the worlds of the mundane and the sublime. Virtually all primitive religions invoked divine intervention in human affairs, and the earliest philosophers pondered the metaphysical aspects of their existence at least as much as the prosaic. Even cursory study of such relics, writings, and rituals suggests a synergistic triad of motivations for the extraordinary efforts expended in such activities:

1) practical benefits in daily life;
2) satisfaction of spiritual yearnings;
3) intellectual curiosity about the nature of the supernatural.

In one form or another, these same three basic desires have sustained major commitments to this component of human experience throughout all subsequent periods of history.

Certainly the mystical aspects of early man's activities did not evaporate with his assembly of more complex societies; they did, however, become more structured and regimented. The classic civilizations of Egypt, Greece, Rome, and of the Near and Far East dealt extensively and systematically in mystical and magical techniques for a variety of pragmatic, spiritual, and academic purposes. The Delphic Oracle, like others of its kind, was politically important from earliest times to the age of Alexander the Great and was routinely consulted on problems as diverse as medical epidemics, the constitutions of the Greek city-states, and the propitious location of new colonies. The roles of religious temples and celestial observatories were intertwined in the search for extraterrestrial influences on agriculture, military operations, public health, politics, birth and death. While priesthoods established procedural recipes for personal spiritual conduct, the religious and secular philosophers of both East and West attempted to bring intel-

Delphic Oracle Consulted by King Aegeus (Greek Vase, ca. 400 B.C.)

lectual order and formalism to the relationship between mystical experience and common activity. Despite their varied geographic and temporal origins, such major ontological manuals as the Chinese *I Ching*, the Indian *Vedas* and *Upanishads*, the Egyptian and Tibetan *Books of the Dead*, the Hebrew *Pentateuch*, and many others shared the common purpose of addressing man's individual and collective communion with the essential forces of life and of the Cosmos.

These intertwined cultural patterns of mystical pragmatism, worship, and scholarly contemplation persisted into the intellectual gloom of the European Dark and Middle Ages, and back out again into the Renaissance and Enlightenment. Comparable patterns can be traced in Eastern societies and through more isolated cultures of the Southern Hemisphere. Even in more modern eras, despite the pervasive growth of technologies, the ascendancy of materialistic philosophy and analytical logic, and the continuous increase in intellectual sophistication, various ramifications of this same basic hunger for mystical participation in the metaphysical realms of experience continued to assert themselves in only slightly altered versions of their ancestral formats.

Nowhere has the interplay of the three motivations for mystical involvement been more intense than in the development of organized religions, and nowhere have anomalies played a more central role. The dominant dogmas of world history have invariably claimed their original authority from certain constellations of prophecy and miraculous events, and even contemporary formalisms and literature remain thoroughly laced with mystical processes and mythical allusions. The Judeo-Christian tradition, like virtually all other enduring forms of religious doctrine, has continued to rest on presumptions of divine power capable of transcending ordinary physical processes to achieve tangible benefits. The Bible, along with other basic theological documents, recounts a broad range of miraculous interventions in such matter-of-fact tones that one is inclined to believe that people of its times and places accepted reports of such events rather routinely and were willing to rely

on such mechanisms in their practical as well as spiritual lives. In fact, the Bible is an excellent catalogue of anomalous phenomena; every category of consciousness-related anomaly we shall later be discussing, and many others as well, is illustrated there in one form or another.

EXODUS CHAP. VII.
Aarons rod devours ỹ Magicians rods.
30
EXODUS 7.Verſe 12.
For they caſt down every man his rod, and they became ſerpents: but Aarons rod ſwallowed up their rods.
170.

Alongside the enduring testimony of established religions to the importance of mystical dimensions in human life runs a less coherent, but equally intense heritage of underground occult organizations and traditions offering alternative modes of expression of the same basic desires. Although less publicly structured or condoned, the genre of supernatural power epitomized by Macbeth's witches has been practiced, sought, and feared throughout history, and the same categories of anomalous phenomena have been featured therein.

Nor has mystical activity been restricted to religious, counter-religious, or other group-ceremonial formats. Individual propensities to mystical practice have ever taken as broad a variety of expression as any other aspects of the personality. Rare is the individual who has not developed some pattern of superstitious behavior or spiritual rumination that reveals a latent or active suspicion of metaphysical reality. Whether avoiding the Ides of March or cracks in the sidewalk, consulting a horoscope or an interpreter of dreams, or imploring an explicit or nameless God in time of peril, man continues to display his personal need for cosmic contact.

Despite its sophistication, our contemporary life has retained all of these collective and individual mystical characteristics. Religious observances still feature prayer, incantation, and sacrament; lodges, orders, and cults continue to thrive on secret rituals explicitly relating them to ancient mystical traditions; spiritualist movements purporting communication with disembodied souls flourish in many cultures; and Eastern spiritual disciplines of meditation, yoga, and the martial arts have propagated into pragmatic Western societies. On one side of the book stores, the occult shelves bulge with latter-day ghost stories, manuals on astrology, witchcraft, and the Tarot, and do-it-yourself guides to developing psychic powers, while on the other side, innumerable handbooks on self-actualization and enhanced performance continue to sell. And, however shallow its present implications, "In God We Trust" remains emblazoned on our currency, along with the classic Hermetic symbol of the eye and pyramid. Such persistence of mystical custom, and indeed the very tendency to interpret much historical evidence in predominantly metaphysical terms, betrays the depth and intensity of this dimension of the modern human psyche,

"As a practical system, Magic is concerned not so much with analysis as with bringing into operation the creative and intuitive parts of man."

ISRAEL REGARDIE*

* Sources for sidebar quotations are listed at the back of the book.

despite its superficial subjugation to contemporary deterministic cultural styles. As from his beginning, man continues to seek the power to influence his own destiny by mystical means; to yearn for profound transcendental experiences; and to search for comprehension of the meaning and mechanics of those processes.

Although it has labored to disclaim such intangible numinous influences, at its deeper levels even modern science cannot maintain total immunity to metaphysical issues. Again and again, in numerous sectors of its current activity, it reaches impasses for its present paradigm where some relationship between consciousness and the expression of physical reality must objectively be confronted. It is in this particular sense that the work described in this book modestly takes its place in the long mystical line.

Tarot Card: "The Star"

3

SCHOLARLY STREAM

The first requisite for any new piece of scholarly research is full assessment of all foregoing work that is directly pertinent to the proposed project. For the difficult expedition contemplated here, such preparation is especially important to gather all possible prior insight, to preclude repetition of errors, and to plan the most propitious agenda and incisive strategies. However, it quickly becomes evident that any attempt to review systematically all scholarly study of the role of consciousness in physical processes faces three major impediments that are themselves important indicators of the character and implications of this topic. First, the stream of such scholarly effort is not at all well circumscribed, but flows over virtually all recorded history, all cultures, and all intellectual categories. Second, the subject itself is so intrinsically holistic that any conceptual dissection along disciplinary lines tends to obscure rather than to illuminate the phenomena. And third, even if such dissection were feasible in the abstract, the discipline-segregated, coldly analytical style of thought and language that characterizes most modern scholarship has prevailed for only a relatively brief period of academic history. Over the larger portion, wherein much relevant study of this topic has been performed, philosophy, theology, medicine, art, and science have been tightly intertwined, and purely intellectual attention to any topic, most especially this one, has been largely inseparable from closely related aesthetic, metaphysical, and, in the deepest sense of the term, magical concerns. Only since the so-called "Scientific Revolution" of the seventeenth century, when the academic boundaries of intellectual endeavor began to be more sharply drawn, can a specialized body of strictly scholarly work—in the contemporary sense of the term—on consciousness-related anomalies be reasonably catalogued. These difficulties notwithstanding, even a cursory survey of the full pattern of scientific and philosophical evolution reveals the immense importance of metaphysical concepts and values in the study of many topics, over many ages.

The trail of Western scientific thought can be clearly traced only back to the philosophers of ancient Greece. Prior to that time, several millennia of substantial scholarly activity in Egypt, Babylonia, India, and the Orient tend to be veiled in clouds of

Thoth: Egyptian Symbol of Scientific Insight

myth and legend wherein the worlds of men and their gods are virtually indistinguishable. In testaments of uncertain origin, we are told of superhuman heroes, such as the Indian Manu, the Greek Prometheus, the Biblical Seth, the Egyptian Thoth, or the legendary progenitor of the Hermetic alchemical tradition, Hermes Trismegistus, who imparted to man divine gifts of knowledge of the secrets of Nature. In such writings stand the earliest references to an array of mystical principles that clearly influenced Greek science, and thereby the course of all subsequent Western scholarship. The most fundamental of these dealt with the analogical relationship between macrocosm and microcosm:

> What is below is like that which is above, and what is above is like that which is below, to accomplish the miracles of one thing.[2]

The first well-documented Greek scientific explorations, such as those of Pythagoras in the sixth century B.C., were characterized by a similar metaphysical conceptualization of a living natural world in which all things and all processes were permeated by some form of mind or spirit, and thus were microscopic reflections of the divine macrocosm. Mathematics and astronomy were regarded as the language of that divine mind—the sacred tools whereby its mysteries might be penetrated and its forces invoked to practical purposes. Following in this line, Socrates championed the mind of man as the ultimate subject of study and exhorted his students to understand the world by understanding themselves. The greatest of those pupils, Plato, escalated these concepts to insist on the primacy of pure reason, or mind, over the material senses, or matter. In contrast, Plato's most eminent student, Aristotle, while acknowledging his mentor's emphasis on the abstract or magical qualities of experience, favored a more causal and deductive approach to knowledge that was to become the backbone of science as we now define it. Nonetheless, Aristotle maintained his belief that all natural processes were driven by some divine purpose to achieve their own forms of perfection, and that the particular form taken by any substance was imposed on it by mind.

The same division of interpretation that distinguished the Platonic and Aristotelian schools has, in one form or another, been a principal factor in all subsequent Western thought. Debates over the relative primacy of mind versus matter, form versus substance, or psyche versus physics, continue to this day in such contexts as the "mind/body problem," the "wave/particle duality," or the "free will/determinism" dichotomy. The balances of favor have shifted back and forth over times, cultures, and topical areas, but almost every major thinker, sooner or later, explicitly or implicitly, has been drawn into confrontation with this fundamental question.

"There is no longer a dualism of mind and matter, but of waves and particles; these seem to be the direct, although almost unrecognizable, descendants of the older mind and matter, the waves replacing mind and the particles matter."

JAMES JEANS

The Romans were better soldiers than scholars, better technologists than scientists, and better politicians than philosophers, and contributed little more to this dialogue beyond some preservation of the Greek records. After the fall of their Empire, the only agencies preserving Europe from total barbarism were the diffused Roman Church and the remote Byzantine culture. Virtually all intellectual activity of the Dark Ages was concentrated in isolated Christian monasteries, where even the most learned monks knew only simple arithmetic and the rudiments of Greek astronomy. For several centuries, little distinction was drawn between mathematics and astrology, or between science and alchemy, all of which remained dominated by a theological doctrine that imposed its own metaphysical themes. The study of nature was pursued primarily for its analogy to prevailing religious belief, and scientific truth was valued only insofar as it provided evidence of divine activity.

During this European dark period, many brilliant Arab scholars were studying and translating the substantial portion of Aristotle's writings preserved in Byzantine libraries. The ideas gleaned from these were integrated with other metaphysical and scientific concepts drawn from numerous Indian and Chinese mathematicians, astronomers, physicians, and alchemists, and the composite insights generated in this fashion eventually found their way into Latin to be debated in ecclesiastical terms by medieval Christian intellectuals. Initial reactions to the writings of such "heathen" luminaries as Al-Kindi, Avicenna, Averroes, and Maimonides were highly skeptical, if not hostile, but through the persistent efforts of visionaries like Albertus Magnus, the great philosopher, alchemist, and teacher of Thomas Aquinas, some Aristotelian cosmology and metaphysics were gradually assimilated into the prevailing church doctrines, allowing a cautious return to creative scholarship in the Christian world.

Yet throughout this long period, astronomy, astrology, mathematics, philosophy, alchemy, and medicine were still regarded as aspects of a single science heavily imbued with Hermetic mysticism that could be tolerated only insofar as it reinforced pontifical principles. Illustrative of the scholar's dilemma was the fate of the thirteenth-century Franciscan monk, Roger

An Hermetic Philosopher

Bacon, who performed remarkably rigorous experiments in optics, chemistry, and mechanics, presaging the invention of eyeglasses, telescopes, gun powder, and steam engines, but spent twenty-two years in prison on accusations of practicing magic and wizardry, and for discussing such sacrilegious possibilities as:

> . . . instruments of wonderfully excellent usefulness, such as machines for flying, or for moving in vehicles without animals and yet with incomparable speed, or of navigating without oarsmen more swiftly than would be thought possible through the hands of men. For these things have been done in our day, lest anyone should ridicule them or be astonished. . . .
>
> Flying machines can be made, and a man sitting in the middle of the machine may revolve some ingenious device by which artificial wings may beat the air in the manner of a flying bird. . . . Also machines can be made for walking in the sea and the rivers, even to the bottom, without danger.[3]

Even well into the seventeenth century, the perils of scientific scholarship had not abated. If anything, they had been exacerbated by the defensive reaction of the Roman Church to the Protestant Reformation on the one hand, and to a revival of Platonic and Hermetic philosophy on the other, both of which threatened those Aristotelian doctrines that had by this time become theologically acceptable. The brilliant astronomer Giordano Bruno, for example, was burned as a heretic in 1600 for disputing the finite circumscription of the universe presumed in the prevailing church dogma, and Galileo faced excommunication for his defense of the Copernican heliocentric model of the solar system.

Copernicus himself, usually represented as having broken through the established geocentric fallacy with the cold weaponry of scientific observation and logic, was strongly influenced by the Hermetic aesthetic. In his own choice of metaphors:

> In the middle of all sits the Sun enthroned. In this most beautiful temple, could we place this luminary in any better position from

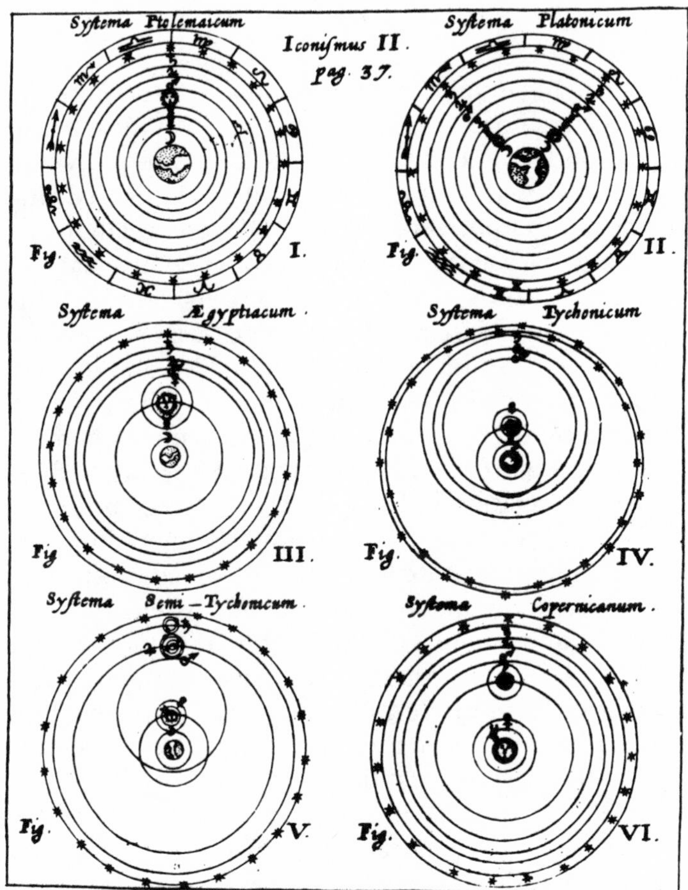

Alternative Models of the Solar System (17th Century)

> which he can illuminate the whole at once? He is rightly called the Lamp, the Mind, the Ruler of the Universe: Hermes Trismegistus names him the Visible God, Sophocles' Electra calls him the All-Seeing. So the Sun sits as upon a royal throne, ruling his children, the planets which circle around him.[4]

One of the greatest champions of the mystical Neoplatonic movement in European intellectual circles was the German physician Theophrastus Bombast von Hohenheim, more popularly known as Paracelsus, as renowned for his contributions to alchemy as for his immense influence on the development of modern medicine. For Paracelsus, as for many other scholars of the sixteenth and seventeenth centuries, the ancient practice of alchemy represented a convergence of the spiritual and physical domains of experience via the Hermetic relationship of microcosm and macrocosm. Such work could be summarized in much the same terms we might use to describe our own: the systematic study of the interaction of consciousness with the physical world; the examination of the tangible ramifications of that interaction; and the use of metaphor in comprehending and representing the process. Although some alchemical practitioners favored a physical interpretation while others preferred a more metaphysical emphasis, it was not until late in the seventeenth century that alchemy diverged into two distinct paths. The physical experiments, divorced from their metaphysical context, mutated into common chemistry; the metaphysical, unconstrained by the scientific discipline of the laboratory, took its own enigmatic, mysterious route, to be rejected by most mechanistic scholars as empty superstition.

"When intersected by a plane, the sphere displays in this section the circle, the genuine image of the created mind, placed in command of the body which it is appointed to rule; and this circle is to the sphere as the human mind is to the Mind Divine. . . ."

JOHANNES KEPLER

It was the earlier, integrated alchemical tradition that underlay the genius of Johannes Kepler, and of his even more mystical mentor, Tycho Brahe. Like Pythagoras, Kepler believed firmly in the mathematical harmony of the universe, which he deemed to have been created by a playful Magician and to be expressed through the abstract symbolism of music and number.

The writings of William Gilbert, the first scholar to render the long-standing empirical evidence of geomagnetism into a scientific framework, reveal a similarly magical component in his scholarly reasoning:

> Aristotle's world would seem to be a monstrous creation, in which all things are perfect, vigorous and animate, while the earth alone, luckless small fraction, is imperfect, dead, inanimate and subject to decay. On the other hand, Hermes, Zoroaster, Orpheus recognize a universal soul. As for us, we deem the whole world animate, and all globes, all stars and this glorious earth too, we hold to be from the beginning by their own destinate souls governed and from them to have the impulse of self-preservation. . . . Wherefore not without reason Thales, as Aristotle reports in his book *De Anima*, declares the lodestone to be animate, a part of the animate mother earth and her beloved offspring.[5]

Even the mighty Isaac Newton, unchallenged father of the mechanistic tradition of our modern science, has properly been described as

> . . . not the first of the age of reason. He was the last of the magicians, the last of the Babylonians and Sumerians, the last great mind which looked out on the visible and intellectual world with the same eyes as those who began to build our intellectual inheritance rather less than 10,000 years ago.[6]

Influenced as much by the alchemical and magical principles of Hermetic philosophy as by the Cartesian rationalism and Baconian methodology of his times, Newton's scientific brilliance was the progeny of the marriage of his mystical and mathematical insights. Counterbalancing his monumental works in classical mechanics, optics, and gravitation, Newton wrote extensively on Biblical prophecy, studied the occult symbolism in the structure of Solomon's Temple and the Great Pyramid of Cheops, and expended immense effort performing experiments in his private alchemical laboratory. His concept of a universal gravitational force connecting all bodies in the cosmos was clearly influenced by ancient mystical ponderings on the subject, and his views on mechanistic causality were cast in terms of some willful, mind-like influence on the course of the physical world. As another biographer has observed, Newton regarded the ultimate mechanism of change in the universe to

The Alchemical Marriage (13th Century)

reside in "the mystery by which mind could control matter."[7]

Only after the acclamation of the "scientific method" around the turn of the seventeenth century did more stringent approaches to scientific and philosophical topics begin to crystallize. The principal architect of this new methodology was the prototypical "Renaissance Man," Francis Bacon, renowned philosopher, scientist, and statesman, whose firm insistence on balance between empirical evidence and deductive logic was to influence all subsequent scientific strategy. Bacon's florid representation of this essential dialogue between experiment and theory closely parallels our own two-step metaphor:

> . . . Those who have treated the sciences were either empirics or rationalists. The empirics, like ants, only lay up stores, and use them; the rationalists, like spiders, spin webs out of themselves; but the bee takes a middle course, gathering her matter from the flowers of the field and garden, and digesting and preparing it by

her native powers. In like manner, that is the true office and work of philosophy, which, not trusting too much to the faculties of the mind, does not lay up the matter, afforded by natural history and mechanical experience, entire or unfashioned, in the memory, but treasures it, after being first elaborated and digested in the understanding; and, therefore, we have a good ground of hope, from the close and strict union of the experimental and rational faculty, which have not hitherto been united.[8]

Particularly illustrative of this change in scientific tenor was the formation in 1662 of the Royal Society of London, whose purpose the physicist Robert Hooke defined in terms that specifically excluded the less tangible dimensions of inquiry:

> To improve the knowledge of naturall things, and all useful Arts, Manufactures, Mechanick practises, Engines and Inventions by Experiments—(not meddling with Divinity, Metaphysics, Moralls, Politicks, Grammar, Rhetoric, or Logick).[9]

Notwithstanding this preoccupation with scientific rigor in general, and Hooke's admonition in particular, many major research figures of this era maintained substantial interest in various esoteric phenomena. Indeed, the new commitment to scientific determinism cast such anomalies into much sharper relief, in the sense that they now fell clearly outside of the new mechanistic paradigm and thus mounted a direct challenge to it. One of the earliest tasks the Royal Society set for itself was the scientific investigation of such topics as astrology, alchemy, prophecy, magic, and witchcraft, and a number of its most prominent Fellows, who were themselves practicing alchemists and students of the occult, participated in these studies. Bacon himself believed in natural magic and proposed systematic investigation of telepathic dreams, psychic healing, and "experiments touching transmission of spirits and the force of the imagination."[10] Despite his public injunction to the contrary, even Hooke lectured and wrote on keenly metaphysical topics. One of his statements holds considerable relevance to the topic of this book:

> Now though by what I have been saying, I have endeavoured to shew that the Soul has by its Radiation a more than ordinary and commanding Power over all the Ideas placed within the Repository; yet I would not be understood so to limit its Sphere of Radiation, as not to suppose that it may have a much bigger Sphere of influencing Power, and thereby may extend it, not only to all and every Point of the Body enlivened and preserved by it; but possibly it may extend even out of the Body, and that to some considerable Distance, and thereby not only influence other Bodies, but be influenced by them also. And upon this account I could produce a Multitude of Observations and Reasons, to prove not only the Possibility, but the Probability, nay almost Certainty of such an Influence, and this from the Sensibleness of other Ideas, . . .[11]

Robert Boyle, pioneer in the kinetic theory of gases and lifelong friend and colleague of Isaac Newton, shared with him an intense interest in alchemy and mysticism. Despite being chastised by scientific colleagues for over-publicizing such activities, he nonetheless persisted in his commitment to the Hermetic heritage:

> But, however, since I find myself now grown old, I think it time to comply with my former intentions to leave a kind of Hermetic legacy to the studious disciples of that art, and to deliver candidly, in the annexed paper, some processes chemical and medicinal, that are less simple and plain than those barely luciferous ones I have been wont to affect, and of a more difficult and elaborate kind than those I have hitherto published, and more of kin to the noblest Hermetic secrets, or, as Helmont styles them, *arcana majora*.[12]

Some historians of science go so far as to contend that the Royal Society was actually founded as a vehicle to incorporate Hermetic, Neoplatonic, and Rosicrucian concepts and heritage into scholarly research, while protecting the participants from religious persecution. In this view, the careful public wordings of Hooke, Boyle, Newton, and others are interpreted as guises to throw religious authorities off the scent by claiming strictly

A Masonic Medal Reflecting the Enduring Hermetic Heritage (18th Century)

empirical approaches and ostensibly eschewing heretical ideas.[13]

During the same period, a number of less formal intellectual circles also met regularly to discuss a variety of metaphysical topics. One group included another close friend of Newton, the philosopher Henry More, and the theologian Joseph Glanvill, both Fellows of the Royal Society. The substance of their investigations was published in a number of books under such intriguing titles as *Saducismus Triumphatus or, Full and Plain Evidence Concerning Witches and Apparitions*; *Immortality of the Soul*; and *An Antidote Against Atheism*, wherein More explained "the parallel economies of God's and Satan's earthly vicegerents."[14]

This lineage of Western scholarship could, of course, be substantially buttressed by comparable lists from the heritage of the Near and Far East or from the tribal traditions of less sophisticated cultures where sensitivity to mystical dimensions and their integration into intellectual practice were, if anything, yet more extensive. Such monumental ancient tomes as the *Pyramid Texts*, the *Sefer HaZohar*, the *Yoga Sutras*, or the *Avesta*, comprise remarkable admixtures of pragmatic philosophy and spiritual insight whose timeless relevance has not yet been fully assimilated, despite countless interpretive efforts by many subsequent cultures and ages. Certainly the historical influence and conceptual utility of the Oriental perspectives on modern Western science and technology has become progressively more evident.

Nor would it be fair to contend that the mystical dimensions of scientific creativity were uniformly acknowledged by all scholars of all eras prior to the Enlightenment. Numerous examples could be cited where materialistic and mechanistic logic were dominant, or at least sharply segregated from the aesthetic and intuitive components of the scholar's activity: Democritus, the Greek founder of atomic theory, largely eschewed spiritual argument and contended intensely with the views of Heraclitus on such issues; the eclectic genius Leonardo da Vinci, despite his extraordinary artistic sensitivity, maintained that physical phenomena should be explained solely in terms of natural causes; and the mathematician and philosopher René Descartes, although motivated by spiritual concerns, relied almost totally on rational and reductionistic logic for his highly influential theorems. On balance, however, the historical message remains clear: any specific inquiry regarding the course of disciplined scientific attention to consciousness-related phenomena leads to the more fundamental recognition that over most of the full lineage of Western and Eastern scholarship, mystical and metaphysical perspectives have been integral components of scientific study in general and have substantially influenced its premises, methods, and conclusions.

The principles publicly and collectively espoused by the founders of the Royal Society concerning separation of science from theology also stimulated and guided a gradual division of the heretofore confluent scholarly stream into the many disciplinary branches we now number in modern intellectual endeavor. The first major schism of this sort appeared between the classical academies of science and philosophy that prior to that time had been inextricably intermingled throughout many eras, topics, and personalities. Like Janus, science and philosophy had originated and long endured as two faces of the same scholarly search, the one looking for the realities of nature, the other for the nature of reality, in either view combining the same empirical experience with analytical conceptualization to achieve incremental knowledge. Certainly, such dominant figures as Aristotle, Copernicus, Roger and Francis Bacon, Paracelsus, Kepler, Newton, and Descartes could as well be labeled in either category on the basis of the persistence of their scholarly influence into both of the modern, separated domains. But as mechanics and rationalism began to replace alchemy and metaphysics as the dominant strategies of analytical endeavor, science and philosophy broke into individual streams flowing along their distinctive, albeit parallel courses, and our review has little choice but to adopt the same strategy from this point.

Over its ensuing three centuries of independent existence, the field of philosophy has generated vast arrays of hyperarticulated theses that defy systematic survey. Nonetheless, even the most cursory review of this immense reservoir of contemplation confirms that the mind/matter, spirit/substance, subject/object dichotomy has remained very close to the core of most philosophical thought. A full spectrum of persuasions may be found on this issue, ranging from the objective materialism of Bacon, Hobbes, and the English empiricists, along with Voltaire and the French *philosophes*, on one extreme, to the more subjective Neoplatonic approach of Berkeley, Rousseau, Hegel, and the German idealists, along with the American transcendentalists, on the other. The centroid of opinion has shifted back and forth over time in response to prevailing social and political trends and the compounding of scientific knowledge, but in every era efforts have been made to establish dialectical combinations of these two perspectives in a tolerant and symbiotic fashion. Of the many philosophers who have approached the mind/matter question from such integrated viewpoints, three have had particular impact on our own orientation: Immanuel Kant, Henri Bergson, and William James.

Kant, eminent thinker of eighteenth-century Germany, aspired to a form of critical philosophy that would synthesize the rational skepticism of David Hume with the romantic idealism of Jean-Jacques Rousseau. He distinguished a perceived world of empirical sensations, or "phenomena," from an objective world of transcendental, unknowable things-in-themselves, or "noumena," and proposed that experience was the product of the mind, with its inherent organizational categories, working on the raw input of the senses. To Kant, time, space, and other "laws of nature," including causality, were products of the human mind rather than objective realities of a physical world:

> We ourselves introduce that order and regularity in the appearance which we entitle "nature." We could never find them in appearances had we not ourselves, by the nature of our own mind, originally set them there.[15]

A century later, protesting the materialistic positivism that attended the rapid industrialization of Europe, the popular French philosopher Henri Bergson also became an eloquent spokesman for the active role of mind and free will in the creation of reality. Posing the question in terms of a fundamental tension between the life force, which he termed "élan vital," and the "inertia" of the material world, Bergson maintained that consciousness must be coextensive with life itself. "Consciousness," he observed, "seems proportionate to the living being's power of choice. It lights up the zone of potentialities that surrounds the act."[16] While he acknowledged the connection between the processes of consciousness and the physical brain, he insisted that "the brain is no more coextensive with consciousness than the edge is with the knife."[17] Man could attempt to understand or describe the physical world via his intellect, but he could only experience it directly through intuition; hence, as Kant had claimed, time and space must be subjective properties of the intellect. Bergson's writings exerted a powerful impact on scholars of many fields, including the prominent paleontologist and theologian Pierre Teilhard de Chardin, who incorporated their premises into his own philosophy of consciousness-driven evolution, and the quantum physicist Louis de Broglie, who advocated a metaphysical dimension to complement reductionistic science.

Yet perhaps the most eloquent commentator on both the epistemology and the mechanics of consciousness was the American philosopher and psychologist William James, who repeatedly stressed the importance of incorporating anomalous psychic experiences in any comprehensive assessment of human nature. Throughout his writings on philosophy, psychology, religion, and psychical research runs the common theme of consciousness creating its reality by selectively sorting out from the "blooming, buzzing, confusion" of environmental stimuli those portions pragmatically useful to its purposes or otherwise reinforcing the expression of its will.[18] James's ideas pervade much of the subsequent history of psychic research, as well as our own work, and will appear later in other portions of this text, but for the moment we must set him aside, along with many other potentially relevant philosophers, to trace a few remaining tributaries of the separated scholarly stream.

The intellectual Enlightenment that followed the Scientific Revolution served to dispel much of the long-standing tension between secular research and religious doctrine. Centuries of public hysteria and religious paranoia over sorcery and witchcraft, manifested in a tragic sequence of trials, tortures, and executions, had begun to subside, and religious scholars were

Witchcraft on Trial (19th Century)

again able to address supernatural phenomena with some dispassion. In the mid-eighteenth century, the Roman Church authorized Cardinal Prospero Lambertini, who later became Pope Benedict XIV, to carry out a scholarly investigation of reported psychic events to distinguish authentic miracles appropriate to cases of beatification. His conclusions, reported in *De Beatificatione Servorum Dei et De Canonizatione Beatorum*, were remarkably unecclesiastical and substantially presaged much later comprehension of such topics: namely, that the grace of prophecy, "whereby a man may know. . . . future, past, or distant or hidden present things, or the secrets of hearts, or inward thoughts" 1) was not necessarily a divine or demonic miracle, but could have natural origins; 2) was more likely to occur to illiterate persons than to trained minds busy with "internal passions or external occupations"; 3) occurred more often in sleep than in waking; 4) was often indistinguishable from personal thoughts; 5) frequently took symbolic forms; and 6) could occur in heathens, children, women, animals, and fish, as well as in "holy people."[19]

During this same period, the newly emancipated secular study of mystical affairs also attracted a number of outstanding scholars from many disciplines, none greater than Emanuel Swedenborg—mathematician, engineer, psychologist, physicist, physiologist, inventor, member of the Swedish Parliament, musician, and clairvoyant. Asserting that "now it is allowable to enter intellectually into the mysteries of faith," Swedenborg propounded a philosophy wherein the relationship between the natural world and the human mind was established through a "Doctrine of Correspondences," and whereby "love or will is the very life of man."[20] As one of the most profound mystical thinkers of his time, his ideas permeated much contemporary and subsequent literature and influenced many later intellectuals, including Kant, Bergson, and William James.

From the medical profession of the eighteenth century emerged another pertinent contribution to the scholarly stream of anomalies research in the work of the physician Anton Mesmer on hypnotic therapy. Mesmer proposed that all things in nature radiated a cosmic vitality, or "animal magnetism," the imbalance of which he deemed responsible for many kinds of disease.[21] He maintained that when rapport was established between patient and healer, this influence circulated through the two as if they were interconnected, resulting in reestablishment of magnetic equilibrium and facilitating healing. Mesmer's reports of such miraculous cures were met with ridicule by his contemporaries, who accused him of charlatanism when he could not describe the process adequately. However speculative Mesmer's concepts may have been, they laid the foundation for later investigations of hypnosis by others, including Sigmund Freud and his followers.

Near the turn of the twentieth century, Freud's pioneering efforts in clinical psychology and psychoanalytic theory opened deeper comprehension of the capacities of human consciousness, and introduced a new conceptual framework and vocabulary for representation of extraordinary effects. Freud's recognition and exploration of the unconscious mind and of the function of dreams supported the view that internal mental processes, rather than external influences, underlay much of perceived reality. His former protégé, Carl Jung, examined the unconscious mind in much greater depth, where he encountered a vast assortment of anomalous phenomena that formed the basis for numerous comprehensive studies of mystical and occult topics. Jung began his professional career in 1902 with a doctoral thesis entitled "On the Psychology of So-called Occult Phenomena" and went on to write extensively on dream symbolism, religious and transformational experience, mythology, Eastern religion and philosophy, alchemy, belief in spirits and the immortality of the soul, flying saucers as a modern myth, and various other mystical and psychical phenomena. His concepts of the collective unconscious and its archetypes have since become central themes in some forms of modern clinical psychology. Of special relevance is his theory of an "acausal connecting principle," or "synchronicity," postulated to be responsible for apparently random coincidences of meaningful events.[22] This work was published in collabo-

An Acausal Connecting Principle (Jacob's Ladder, *by William Blake–18th Century)*

ration with the physicist Wolfgang Pauli, and clearly influenced the subsequent evolution of both careers and contributed fundamental insights to both psychology and physics.

As the rapidly proliferating and segregating Western sciences and philosophies crystallized throughout the nineteenth century and expanded into their modern eras, new technical options for addressing the role of consciousness in physical reality became similarly abundant, but the increasing insularity and secularity of these disciplines inhibited effective study of this inherently holistic, substantially metaphysical topic. In a few specific venues, some of which are described in the following chapters, the consciousness issue has been sufficiently evident or unavoidable that deliberate consideration of it has been awarded. But in most other disciplines, this dimension has largely been swept under the rug, relegated to an epiphenomenon of the physical brain, or totally rejected. Minor exceptions could be found in several other fields we have not mentioned, but little would be added to our perspective by reciting these. The historical scholarly pattern is already unmistakable: far beyond its strictly intellectual capacities, consciousness possesses and utilizes mystical and metaphysical dimensions for its conceptualization and interpretation of reality. What remains to be determined is the degree to which these participate in the establishment of that same reality.

4

PARAPSYCHOLOGICAL PERSPECTIVE

The first substantial attempt to institutionalize the scientific study of anomalous consciousness phenomena, *per se*, dates to the establishment in London in 1882 of the Society for Psychical Research (SPR). The widespread popularity of spiritualism, with its attendant séances and "table-tippings," prompted a number of notable scholars from widely disparate disciplines, led by the physicist William F. Barrett, the philosopher Henry Sidgwick, the essayist Frederick W. H. Myers, and the psychologist Edmund Gurney, to attempt to address these issues with the same scientific rigor that characterized other current intellectual pursuits. A few years later, a counterpart organization, the American Society for Psychical Research (ASPR), was founded in Boston by several equally distinguished scientists and philosophers, under the leadership of Simon Newcomb and William James. The explicit and implicit charters of these two organizations were nicely summarized by James's proposition:

> Any one with a healthy sense for evidence, a sense not methodically blunted by the sectarianism of "Science," ought now, it seems to me, to feel that exalted sensibilities and memories, veridical phantasms, haunted houses, trances with supernormal faculty, and even experimental thought-transference, are natural kinds of phenomena which ought, just like other natural events, to be followed up with scientific curiosity.[23]

Victorian Table-Tipping

Although these societies attracted a barrage of criticism from the established scientific and intellectual communities, they also attracted the interest and participation of many eminent scholars of the day. Numbered among their presidents were three Nobel laureates, nine Fellows of the Royal Society, one prime minister, an archbishop of Canterbury, and a substantial list of prominent academics including, in addition to the founders, Arthur, Earl of Balfour; Henri Bergson; Sir William Crookes; Sir Oliver Lodge; William McDougall; Gardner Murphy; Gilbert Murray; John, Lord Rayleigh; Charles Richet; Eleanor Balfour Sidgwick; Sir J. J. Thomson (vice president); and G. N. M. Tyrell. Beyond this distinguished roster of officers, such notables as Marie and Pierre Curie; Sigmund Freud; Carl Jung; Alfred, Lord Tennyson; Mark Twain; Alfred Russel Wallace; and William Butler Yeats were drawn into the societies' activities.

Early in this formative period, Myers compiled his many years of personal study in a classic two-volume work entitled *Human Personality and the Survival of Bodily Death*, which he himself described as

> an exposition rather than a proof [of] the thesis that a communication can take place from mind to mind without the agency of the recognised organs of sense. We found that this agency, discernible even on trivial occasions by suitable experiment, seemed

to connect itself with an agency more intense, or at any rate more recognisable, which operated at moments of crisis or at the hour of death.[24]

This monumental tome had drawn heavily from a similarly imposing effort called *Phantasms of the Living* in which Gurney had catalogued a substantial data base of reports of spontaneous supernatural experiences, drawn from fifteen years accumulated research in the SPR publications and files, as well as from a variety of other historical documents, including the literature of witchcraft.

Much of the scientific interest in the topic during this period had been stimulated by a program of careful experiments performed several years earlier by the physicist William Crookes, whom Myers described as "the first leading man of science who seriously endeavoured to test the alleged mutual influence and interpenetration of the spiritual world and our own by experiments of scientific precision."[25] Crookes, discoverer of the element thallium and pioneer in the study of electrical

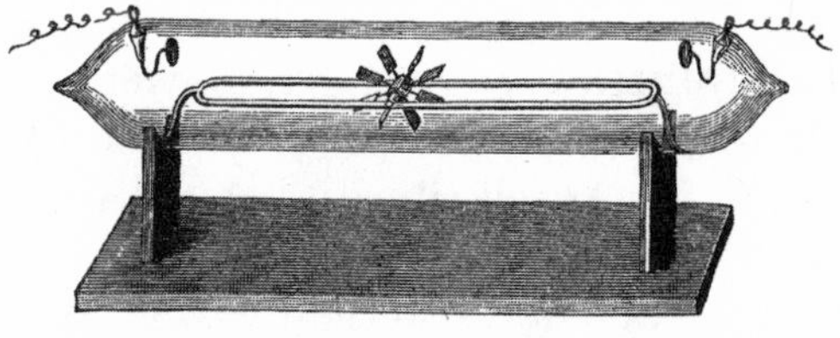

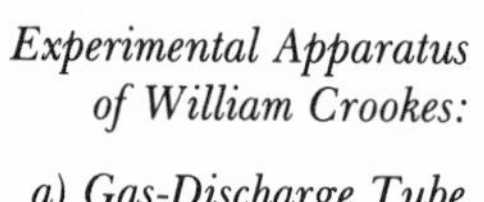

Experimental Apparatus of William Crookes:

a) Gas-Discharge Tube

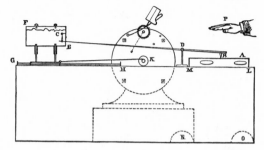

b) Device to Measure Psychic Force

discharges in low-pressure gases, had conducted extensive experiments in psychic communication and physical anomalies with the well-known mediums D. D. Home and Florence Cook, using ingenious instrumentation to detect effects and control against artifacts. Beyond his technical reports, the accounts of Crookes's broadside professional and personal battles with the prevailing scientific establishment make highly entertaining and enlightening reading in their own right.

The commitment to such controversial research of another notable physicist of this period, Oliver Lodge, was perhaps even more intense, more enduring, and more ridiculed by the scientific establishment. A man of deep religious convictions, Lodge progressed steadily from an early skeptical study of telepathy to eventual leadership of the spiritualist movement in England during the First World War. He went so far as to attempt to relate the mechanics of communication with the dead with the physical ether of empty space which was at that time a major enigma in electromagnetic theory, and wrote extensively on such topics as *The Immortality of the Soul* and *The Survival of Man.* Some of Lodge's opinions were sufficiently extreme to attract criticism even from the more conservative members of the SPR.

The Parisian physiologist, Charles Richet, played a seminal role in the early development of this field by introducing concepts of probability and statistical inference into the design and interpretation of experiments in telepathy or, as he termed it, "La suggestion mentale."[26] Although rudiments of probability and statistical theory had previously been applied in other scientific contexts, it is fair to say that Richet's clever implementation of such concepts in his own experimental programs strongly influenced the subsequent analytical strategies of not only parapsychology, but classical psychology, biology, medicine, and even statistics itself.

Although he was a member of both the British and American Societies, Sigmund Freud remained an ambivalent skeptic on the subject throughout his career. On the one hand, in an article entitled "Dreams and Telepathy," he expressed concern lest the reader infer that he was "secretly inclined to support the reality of telepathy in the occult sense."[27] Yet, in a letter to Hereward Carrington, he conceded: "If I had my life to live over again I should devote myself to psychical research rather than to psychoanalysis."[28] He subsequently retracted the statement.

Despite the number of academics involved in the formation and early investigations of the two Societies, psychic research did not prove attractive to the established academic community, and fully articulated university-based research programs were slow to germinate. Over the first quarter of the twentieth century, only a handful of significant academic research efforts were mounted, all of which were performed in departments of psychology. Endowments contributed by William James at Harvard and by Thomas Stanford at Stanford University supported the earliest controlled experiments in telepathic communication in the United States. These studies, along with comparable work at the University of Groningen in Holland, established some methodological precedents and standards of experimental rigor for future research—most notably the concept of randomized experimental design—but failed to generate ongoing enthusiasm for the topic. Much of the academic aversion can probably be traced to the continuing preoccupation of many investigators of that day with mediumistic and spiritualistic phenomena, often elicited under less than ideal conditions for scientific assessment. This focus was especially vulnerable to public gullibility and fraudulent exploitation and overlaid an air of sensationalism on the entire field. Perceived interferences with traditional religious prerogatives further discouraged established scholars and institutions from broader attention to the subject.

The first successful effort to bring psychic research into the academic fold began in the late 1920s when William McDougall, who had been James's successor at Harvard, became chairman of the Department of Psychology at Duke University and appointed J. B. and Louisa Rhine to assess the scientific validity of psychical phenomena. The Rhines' early efforts toward systematic study of postmortem survival soon turned to study of a more tractable range of topics generically termed "psi" (ψ) processes or "parapsychology." This name was chosen both to underscore the derivation of their approach from the established strategies, nomenclature, and perspectives of traditional psychology, as well as to emphasize their conviction that the anomalous phenomena under study were associated with the psyches of living persons rather than with disembodied spirits, as had often been presumed in much of the preceding spiritualistically oriented work.

The Rhines distinguished two major categories of effect in their research. One they called "extrasensory perception" (ESP)—the ability to acquire information inaccessible by known sensory channels; the other, "psychokinesis" (PK)—the ability to influence objects or processes in ways inexplicable by known physical means. Typical of their early experiments were long series of attempts on the part of selected individuals to identify randomized sequences of five geometric symbols in a deck of twenty-five "Zener" cards, or to influence the fall of dice beyond chance expectations. These studies evolved into a variety of more complex ESP and PK tasks to assess the correlations between psychic capabilities and psychological characteristics of their numerous participants.

J. B. Rhine summarized the results of the first two decades of their research in a 1967 address to the American Psychological Association:

Rhine's ESP Card Symbols

> The phenomena that were being studied began to show lawful interrelations and even a degree of unity. One by one the major claims, based originally only upon spontaneous human experiences, were subjected to laboratory test and experimentally verified. . . . Certain general characteristics of the psi process became clear during this period. The most revealing of these is the subject's lack of conscious control over any type of psi ability, a characteristic which accounts for its elusive nature. It was new methodological ground, even for psychology. . . . Also, we were surprised to find that psi ability is widespread, probably even a specific human capacity rather than a capability possessed by a few rare individuals as had been the popular belief. Evidence that psi is not linked with illness or abnormality was another welcome advance. . . . By 1951 . . . a healthy young science was emerging.[29]

In the Rhines' laboratory were established most of the basic concepts, protocols, and vocabulary of this new parapsychological discipline, its first extensive and systematic data bases, and its next generation of investigators. Despite enduring skepticism from most of their scientific peers, these efforts set the style for much subsequent research and influenced the attitudes of a number of prominent scientists of the day. Alluding to this work in a volume of essays on physics and epistemology published in 1961, the physicist Wolfgang Pauli observed:

> In recent times there is an empirical parapsychology which claims to employ the approach of exact science and which works with modern experimental methods on the one hand, and modern mathematical statistics on the other. Should the positive results in the still controversial realm of "extrasensory perception" (ESP) finally turn out to be true, this could lead to developments which cannot be envisioned today.[30]

Pauli went on to urge the scientific community to refrain from dismissing the topic on *a priori* philosophical grounds, insisting that the existence of psychic phenomena should be decided via critical empiricism.

It was a natural consequence of the psychological heritage of the Rhines' laboratory and the many subsequent programs spawned by its students and colleagues that parapsychological research should continue to focus on the human participants as "subjects," rather than on the physical aspects of the phenomena. Even today, most of the parapsychological literature addresses correlations of psychic performance with cognitive processes or personality characteristics such as attitude, motivation, creativity, intelligence, cooperation versus competition, or introversion versus extroversion. Other contemporary efforts investigate the effects of hypnosis, sleep, sensory inhibition, meditation, progressive relaxation, or other altered states of consciousness. Consistent with its aspirations to scientific respectability, and in contrast to its more spiritualistic predecessor "psychical research," parapsychology has tended to concentrate mainly on relatively modest psychic tasks that can be performed under controlled laboratory conditions with well-defined protocols for data acquisition and processing, and to stay well

Saducismus Triumphatus:
OR,
Full and Plain EVIDENCE
Concerning
WITCHES
AND
APPARITIONS.
In Two PARTS.
The First treating of their
POSSIBILITY,
The Second of their Real
EXISTENCE.

By *Joseph Glanvil* late Chaplain in Ordinary to his Majesty, and Fellow of the Royal Society.

With a Letter of Dr. *HENRY MORE* on the same Subject.

And an Authentick, but wonderful story of certain *Swedish* Witches; done into English by *Anth. Horneck*, Preacher at the *Savoy*.

LONDON: Printed for *J. Collins* at his Shop under the *Temple*-Church, and *S. Lownds* at his Shop by the *Savoy*-gate, 1681.

One Early Attempt to Catalogue Extreme Phenomena (17th Century)

within the simple statistical models it has helped to define. Notwithstanding, a few efforts have also been made to extend methodical study into more extreme arenas such as survival and reincarnation, poltergeist effects, and trance-mediumship and séance phenomena, while a number of biologically oriented experiments have addressed questions of psychic healing and anomalous animal and plant behavior.

Given its focus on psychological and physiological parameters, much parapsychological research has tended to be burdened by the same broad interpretations and multi-variable dependencies that restrict the incisiveness of conventional psychology. But despite the tenuous nature of many of its findings, the parapsychological approach has provided important indications about subtle features of the phenomena that might well have evaded a more physical orientation. Beyond demonstrating the inherent difficulties in eliciting psychic effects within excessively sterile laboratory environments, it has revealed the important roles of such psychological factors as the participant's attitude toward the phenomena, the relationship between the participant and experimenter, and the subconscious or unconscious components of the processes involved. A number of other aspects have been shown to complicate the strategy and interpretation of the research on the one hand, yet to provide valuable clues to the nature of the phenomena on the other. These include difficulty in replicating results; "decline" or "extinction" effects, where strong results revert to chance expectancies under repeated testing; "psi-missing," where significant effects opposite to those intended are obtained; "displacement," where responses show correlations with targets or intentions adjacent to those assigned; and "experimenter effects," where certain investigators or laboratories seem regularly to obtain higher anomalous yields than others on very similar tasks. Collectively these findings have defined an elusive family of anomalies that defy traditional scientific replicability and causal representation, yet persist in their frequent, albeit unpredictable, appearance.

Beyond the frustrations posed by the phenomena themselves, the fledgling field of parapsychology has been forced to operate under very limited funding, an enduring lack of academic credibility, and unremitting pressure from a vigorously vocal and occasionally unethical critical community. In addition, it has attracted more than its share of exploitation by those who unreasonably extrapolate its meager findings to sensational implications. On the positive side, these stringencies have inclined most of the researchers toward conservative public postures, faithful disclosure of negative or null results, and a strongly self-critical approach to their work. They have also necessitated an admirable degree of self-sufficiency. Denied publication in established professional journals, the Rhines founded the *Journal of Parapsychology* in 1942; excluded from participation in other professional societies, in 1957 they and their colleagues formed the Parapsychological Association, which in 1969 achieved affiliation with the American Association for the Advancement of Science. Today these remain important forums for discussion and dissemination of research results, and along with the SPR, the ASPR, the Parapsychology Foundation, and Parapsychology Sources of Information, provide the most substantial library services and public interfaces available for the field.

On the negative side, this forced insularity has produced a degree of intellectual inbreeding that has limited the range of conceptualization, implementation, and interpretation brought to bear on the research issues and has perpetuated some less than productive protocols and techniques. For example, the heritage of earlier psychical research, coupled with the predominantly psychological perspective, has tended to concentrate a disproportionate effort on searches for "gifted subjects," at the expense of more sophisticated technical strategies. The dependence on such individuals has entailed a vulnerability to their personal abilities, interests, styles, and integrity, and has reinforced the skeptical perception of the field as one dominated by fraudulent practitioners and gullible pseudo scientists. It may also have served to discourage the involvement of less flamboyant, but highly competent participants who shunned the inevitable public visibility, identification as "psychics," and treatment as objects of, rather than as partners in, the research.

Another limitation of the parapsychological approach has been its reliance on protocols and instrumentation that are inherently constrained to relatively insensitive measurements and slow data acquisition rates. Data thus have tended to be too gross in scale or too embroiled in statistical noise to permit meaningful correlations or legitimate conclusions. Only quite recently have protocols and instrumentation been deployed that are capable of sensitive signal-to-noise discrimination, rejection of experimental artifacts, and acquisition of sufficiently large amounts of data under stable conditions that replicable systematic trends can be identified. Much of this recent enhancement of instrumentation and data processing has been provided by the latter-day involvement of a few natural scientists and engineers, who brought with them a broader range of technologies and theoretical perspectives that infused the field with new ideas and skills, and helped shift the focus of research from questions of "who?" to "how?"

"When psi capacities transcend space and time ever so slightly or infrequently, they are revealing fundamental properties of the human mind as a whole. This capacity to intersect with the physical world through ESP and PK is thus a function of the total personality, not of an abstracted, isolated, momentary mental state."

J. B. RHINE

Similar contributions to parapsychology from other academic perspectives, such as anthropology, sociology, philosophy, and medicine, have also been increasing, and currently the field constitutes a paranormal melting pot for a broad range of intellects who share a common interest in the capabilities and potentials of human consciousness. What they all face at present is a far from convincing, far from utilitarian body of bemusing and contradictory evidence that can neither be scientifically catalogued nor categorically ignored as it stands, but surely demands more extensive and incisive research and contemplation.

5

CRITICAL COUNTERPOINT

It is a common adage that a man may be calibrated by the quality of his critics; the same criterion may equally well be applied to any scholarly enterprise or scientific proposition. Rarely in scientific history has a major experimental anomaly or a novel theoretical model been announced that has not precipitated a barrage of skeptical opposition. In most cases, this critical commentary has been objective and valid and has properly served to discredit erroneous data and incorrect concepts, to redirect interpretation into more productive channels, or to refine ideas to greater effectiveness. In some less fortunate situations, the criticism has been encumbered by status quo intransigence and vested interest, parasitic pseudo-scholarship, uninformed and irresponsible bigotry, or *ad hominem* attacks that have illegitimately persecuted scientific pioneers and suppressed their visionary ideas. It is tempting to digress at this point to compound a comprehensive list of examples, both positive and negative, of the role of critics and skeptics in the progress of science, including such prototypical conflicts as those between the Aristotelians and the Platonists, the geocentrics and the heliocentrics, the creationists and the evolutionists, the classical physicists and the quantum mechanicists, and many other lesser set-tos, but it will better serve our purpose to confine such illustration to the topic at hand.

In no case has the pattern of critical commentary been more unrelenting or more widely ranging in its quality and utility

Science versus Mother Goose (19th Century)

than that regarding anomalous consciousness-related phenomena in general, and parapsychology in particular. Historically, the skeptical opposition to claims of such anomalies has run stride-for-stride beside the efforts toward their credible demonstration and comprehension, and has constituted a messy mixture of help and hindrance. The contemporary criticism of psychic research and the rejection of the phenomena it claims to demonstrate comprise an array of generic and specific objections, each of which has some validity and merits some thought in any sincere attempt to comprehend the topic:

1) demonstrable fraud in collection and reporting of data;

2) naïveté of technique, including inadequate controls, faulty equipment, sensory cuing of participants, experimenter biases, selective treatment of data, suppression of negative results, improper statistical methods, and general experimental and theoretical incompetence;

3) lack of meaningful progress over many years of study;

4) poor experimental replicability;

5) sensitivity of results to participants, attitudes, and laboratory ambience;

6) marginal significance of most results compared to chance expectation;

7) elusiveness of effects under skeptical scrutiny;
8) absence of adequate theoretical framework;
9) inconsistency with prevailing scientific paradigm;
10) incompatability with personal belief systems or common sense.

Obviously this list is somewhat redundant and runs a gamut from technical and procedural objections, through phenomenological inconsistencies, to rather categorical and subjective rejections that are difficult to rebut logically or to preclude in any experimental design. In fact, it is probably fair to say that substantial portions of classical and modern research in the physical, biological, and social sciences would be hard pressed to withstand rigorous application of this full list of challenges. Notwithstanding, the precarious heritage and extraordinary implications of psychic research behoove it to address each of the issues frankly and constructively.

"How do scientists treat factual anomalies? What do we do with evidence that challenges a comfortable view of nature's order? . . . Prompt submission to items of contrary evidence is not . . . the usual response of scientists to nature's assaults upon traditional beliefs."

STEPHEN JAY GOULD

Unfortunately, but undeniably, one or another of the first three judgments may properly be applied to much of the historical data base. By its nature, the topic is immensely vulnerable to fraudulent exploitation and naïve gullibility, and such have indeed occurred to a distressing degree, especially in the early mediumistic investigations. It is also true that the topic has attracted a disproportionate share of less than fully competent investigators, and that it presents extraordinary pitfalls for even the most disciplined scholars. The lack of definitive progress toward comprehension of the phenomena should perhaps be qualified by the relatively minute integrated investment of resources in this field compared to many of the more favored areas of modern science, and by the very brief period over which sufficiently sophisticated equipment, data-processing techniques, and physical models have been applied to the task. Nonetheless, legitimate concerns about the ultimate tractability of the field remain.

The more specific objections 4–8 actually merit examination from two opposing points of view: namely, to what extent do such characteristics indeed invalidate the experimental results; or conversely, to what extent might they illuminate the basic nature of the phenomena? Without question, the dominant experimental frustration in this field is the inability to replicate on demand previously observed anomalous effects, not only at other laboratories with other participants, but even in the original facility, using the original participants, under apparently identical experimental circumstances. This contrariety lends itself to a variety of interpretations: the phenomena are a) illusory; b) rare and bizarre chance occurrences beyond any hope of regularization; c) precipitated, at least in part, by psychological and/or physiological factors that are currently beyond experimental control, but which if fully comprehended would conform to established scientific paradigm; or d) inherently statistical, and possibly quantum mechanical, on a macroscopic scale, thus manifesting themselves with only fractional probability on any given occasion. The last two options are not necessarily mutually exclusive, particularly if one takes the doubly statistical point of view that the human population embodies a spectrum of capability for engendering such effects and that any individual may display a highly variable range of personal capacity, depending on the environmental, physiological, and psychological factors prevailing at a given time.

The tendency of many of the anomalous results to fall just marginally outside of normal chance expectation for the statistical system under study also cuts in two directions. On the one hand, it is tempting to suspect that further refinement of the experimental technique or of the statistical processing of the data would bring the results back into the normal fold. On the other hand, it is theoretically consistent and well established empirically for a broad range of less controversial physical, biological, and psychological systems, especially those involving random processes, that marginally deviant results are far more likely to be encountered than grossly deviant ones. In any case, this characteristic mandates precise diagnostic equipment and, perhaps most importantly, the accumulation of very large data bases to allow systematic trends to emerge clearly beyond the statistical noise.

The evasiveness of the phenomena under carefully controlled and observed study may be the most damning scientific criticism of all, or it may reveal an important and illuminating characteristic. The tendency of a given preliminary or anecdotal effect to disappear or diminish when the experiment is tightened or when it is displayed to a jury of skeptical observers obviously casts major doubt on the scientific validity of the results. Yet it also brings to mind a number of other capabilities of consciousness which, while superficially dissimilar, may not be totally irrelevant, notably the processes of aesthetic, intellectual, and functional creativity. Artistic, musical, or literary composition, lofty philosophical thought, pragmatic ingenuity, or romantic courtship are not usually facilitated by rigid constraints or by the presence of a body of unsympathetic observers. Favorable ambience and mood can be important facilitators of such efforts, and little creative achievement is likely to occur in overly sterile or hostile environments—a truth Richard Wagner vividly portrayed to his own critics by the fate

Wagner Admonishing His Critics: "Would you measure by your rules something that is not governed by them? Forget your own guidelines; seek first for the appropriate rules." (Die Meistersinger: *Act I*)

of young Walther attempting his avant-garde "Trial Song" before a truculent audience of Meistersingers. Virtually every creative person preserves some form of retreat or sanctuary, and even the most analytical scientists will concede the role of unstructured imagination in enhancing their technical insights.

As portrayed in our opening metaphor, no scientific mission can progress very far in the absence of a viable theoretical leg, and this field has suffered badly from that void. Many attempts have been made to propose heuristic models or to transpose established physical or psychological formalisms to the task, but so far none has proven broadly serviceable. Again, this failure permits two disparate interpretations: either the topic as a whole is invalid, or all of the important parameters bearing on the phenomena have not yet been acknowledged in the theories.

The final two criticisms regard inconsistency of the phenomena with established scientific and personal paradigms. While these are valid and powerful professional and personal discriminators that properly predicate great caution and discipline in venturing into any poorly understood field, they cannot be allowed total veto authority if new domains of conceptual experience are to be broached. Any such embargo stifles the most precious attribute of human consciousness, the curiosity about the unknown that drives the mind and spirit of man to evolve, rather than merely to replicate. Notwithstanding the risks of rejection, ridicule, and failure, the prevailing structures of belief must continue to be challenged if human intellect is not to stagnate.

Despite certain inconsistencies and shortsightedness in this list of criticisms, serious attempts to generate and interpret experimental data in this field should acknowledge these concerns and be as responsive as possible to them. Certainly, the importance of extreme care and sophistication in data acquisition and processing is unarguable. Beyond that, given the irregularity of the phenomena, their possible dependence upon a broad range of physical, psychological, physiological, and environmental parameters, and their tendency to appear as marginal deviations from normal statistical distributions, experimental designs must incorporate clear discriminators to distinguish legitimate anomalies from mere random fluctuations. Even so, a premium must be placed on the capacity of the experiments for large data base accumulation and processing. And not least of all, some form of theoretical model, however speculative or ad hoc, should counterbalance the experimental program. In these respects and others, our program of research has attempted to obviate the conventional criticisms, without being excessively constrained by them.

Two Early Telescopes, Attributed to Galileo (17th Century)

QUANTUM CLUES

Toward the close of the nineteenth century, and well into the twentieth, the staid and sophisticated world of classical physics became severely discombobulated by a sequence of empirical observations in the atomic domain that were clearly irreconcilable within its established paradigm—true anomalies in every sense of the term. Spectral radiation from incandescent cavities—so-called "black bodies"—digressed catastrophically from the electromagnetic distribution formulae; absorption and emission line spectra from the simplest atomic species could not even be qualitatively rationalized by classical charge radiation theory; the specific heat capacity of numerous solid materials differed drastically from the predictions of statistical thermodynamics; the energy of photoelectrons from an illuminated metallic surface depended not on the intensity of the light, but on its frequency; and beams of particulate electrons displayed wavelike interference patterns.

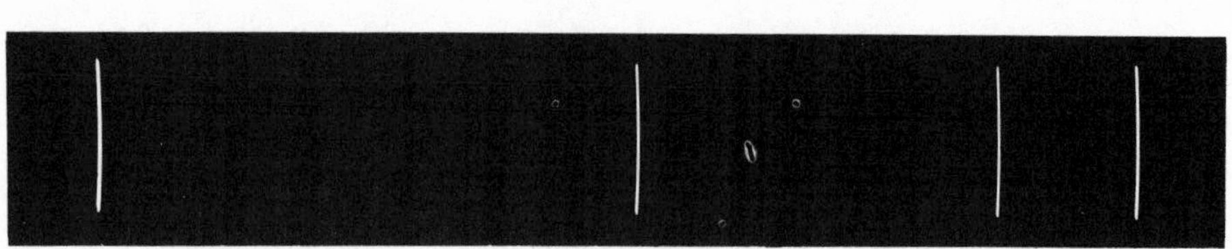

Spectrum of Helium

Much has been written about the revolutionary changes in physical theory that had to be accepted before these anomalous effects could be accommodated, and about the personalities who conceived and implemented such remarkable ideas. The names of Planck and Bohr; of de Broglie, Schrödinger, and Heisenberg; and of Einstein, Pauli, Jordan, von Weizsäcker, and Wigner are revered by students of modern physics for the brilliant manner in which they individually and collectively addressed these challenges. With remarkable insight and no small amount of courage, this group of pioneers hammered out in successively more elegant iterations a new system of physical formalisms based on several radical concepts that were at least as variant with the then prevailing presumptions of classical science as were the empirical anomalies that had provoked them. Electromagnetic energy was postulated to reside in indivisible packets or "quanta," scaled only by the frequency of its oscillation; certain stable atomic configurations were allowed to violate electronic radiation theory; probabilistic aspects were introduced into previously deterministic systems; identical atomic particles were treated as indistinguishable; particles were allowed wavelike properties determined by their mechanical momentum; and the process of observation became consequential in the establishment of physical behavior.

"Once we have decided that the law of causality is by no means a necessary element in the process of human thought, we have made a mental clearance for the approach to the question of its validity in the world of reality."

MAX PLANCK

From a pragmatic standpoint, these efforts were monumentally successful, explicating the backlog of anomalous data with extraordinary precision, and establishing the basis for a grand new era of pure and applied science. But from a conceptual perspective, this compounding empirical success opened a philosophical Pandora's box full of its own anomalous axioms. Immediate and vigorous controversy erupted regarding the logical consequences of such features as the role of the observer in the establishment of reality, the uncertainty principle limitation on the precision of specification of physical properties, and the abrogation of full causality and logical positivism, and many ramifications of these debates endure to this day.

These philosophical implications were certainly not lost on those named pioneers. Yet curiously, the degree to which they and their students pondered and discussed these dilemmas, wrote and lectured about them, seems substantially less appreciated by present practitioners and students of science than are their analytical formalisms and physical data. Nor is it commonly recognized that many eventually were led by such considerations to explore the possible implications of quantum mechanics for various other fields of scholarship substantially removed from the usual purview of modern physics. Niels Bohr maintained a lifelong interest in biology and psychology and speculated about the implications of quantum mechanics for living systems and the applicability of various quantum mechanical concepts as biological metaphors.[31] Wolfgang Pauli's interest in mysticism and the occult led him to explore the influence of archetypal concepts in the development of physical theories.[32] Erwin Schrödinger's study of Eastern and Western philosophy prompted him to write extensively on mystical and metaphysical issues.[33] Werner Heisenberg developed new thrusts in the philosophy of science and its ethical and social applications.[34] Pascual Jordan speculated on the implications of modern physics for the comprehension of various psychic phenomena.[35] Eugene Wigner explicitly acknowledged the need to incorporate consciousness in physical theory.[36] Carl Friedrich von Weizsäcker proposed information as the fundamental currency of physics, and pondered the relationship between Eastern metaphysics and Western science.[37] And Albert Einstein frequently alluded to the bond between science and religion.[38] Beyond these varied personal interests, it is also

"Causality may be considered as a mode of perception by which we reduce our sense impressions to order."

NIELS BOHR

notable that many of this group maintained cognizance of the contemporary parapsychological research then in progress at the Rhines' laboratory and commented upon it.

Through all of their excursions extended the same dichotomy between reductionistic and holistic reasoning that had pervaded the dialogues of Aristotle and Plato, Bacon and Paracelsus, Hobbes and Hegel. In this case the debate was epitomized by the intense instinctive commitment to causal reductionism by Einstein, on the one extreme, versus the empirical holism of Planck, Bohr, and Schrödinger, on the other. To this day, descendants of both schools continue to debate the latest physical anomalies from these alternative perspectives.

Despite their various topical routes and diverse philosophical persuasions, each of the patriarchs eventually came to address

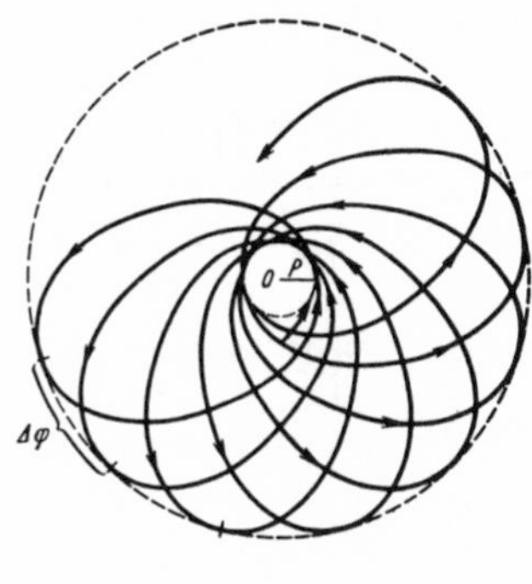

Relativistic Orbits (Precessions Exaggerated):

Electron about the Atomic Nucleus;

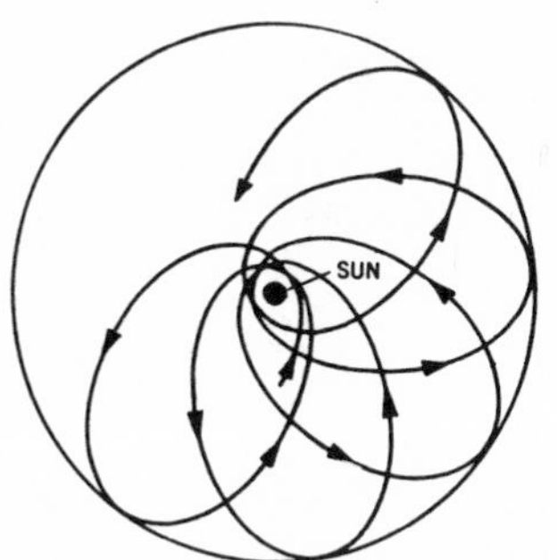

Mercury about the Sun

the ultimate issue of the role of consciousness in the establishment of physical reality. The first of their noble lineage, Max Planck, raised the same critical questions Kant had addressed more than a century earlier:

> All ideas we form of the outer world are ultimately only reflections of our own perceptions. Can we logically set up against our self-consciousness a "Nature" independent of it? Are not all so-called natural laws really nothing more or less than expedient rules with which we associate the run of our perceptions as exactly and conveniently as possible?[39]

He then went on to speak of the importance of retaining a metaphysical component in science:

> . . . scientists have learned that the starting-point of their investigations does not lie solely in the perceptions of the senses, and that science cannot exist without some small portion of metaphysics. Modern Physics impresses us particularly with the truth of the old doctrine which teaches that there are realities existing apart from our sense-perceptions, and that there are problems and conflicts where these realities are of greater value for us than the richest treasures of the world of experience.[40]

At one point Planck explicitly addressed supernatural phenomena:

> Belief in miracle is a very important element in the cultural history of the human race. . . . Is there, in the last analysis, some basically sound foothold for this belief in miracle, no matter how bizarre and illogical may be the outer forms it takes?[41]

A similar logic progresses through Schrödinger's writings:

> It is the same elements that go to compose my mind and the world. This situation is the same for every mind and its world, in spite of the unfathomable abundance of "cross-references" between them. The world is given to me only once, not one existing and one perceived. Subject and object are only one. The barrier between them cannot be said to have broken down as a result of recent

experience in the physical sciences, for this barrier does not exist.[42]

. . . Consciousness is that by which this world first becomes manifest, by which indeed, we can quite calmly say, it first becomes present; that the world *consists* of the elements of consciousness. . . .[43]

I have. . . no hesitation in declaring quite bluntly that the acceptance of a really existing material world, as the explanation of the fact that we all find in the end that we are empirically in the same environment, is mystical and metaphysical.[44]

Mind has erected the objective outside world of the natural philosopher out of its own stuff.[45]

De Broglie also felt keenly about the role of consciousness:

Science is therefore a strange sort of penetration into a world which through human consciousness and reason has learned to become aware of itself.[46]

One of his statements essentially presaged the thesis of this book:

Even in the most exact of all the natural sciences, in Physics, the need for margins of indeterminateness has repeatedly become apparent—a fact which, it seems to us, is worthy of the attention of philosophers, since it may throw a new and illuminating light on the way in which the idealizations formed by our reason become adaptable to Reality.[47]

There is no substitute for more thorough personal reading of the extensive writings of these and the other originators of modern physics on such topics, even though many of these references inexplicably lie dormant or are difficult to access. In the course of developing our own theoretical model, we assembled a compendium of pertinent excerpts from these essays,[48] through which thread a sequence of speculations and inferences that nicely serve to summarize this quantum vector, namely:

1) Physical theory cannot be complete until consciousness is somehow acknowledged as an active element in the establishment of reality.

2) Quantum mechanics has the potential for analogical representation of the interaction of consciousness with its physical environment.

3) The possibility that consciousness can directly influence the behavior of physical systems cannot be categorically dismissed.

4) More extensive and reliable data on the interaction of consciousness with physical systems and processes are required before the two-step scientific process can proceed on this issue.

To our mind, any of these four items would be sufficient motivation for mounting a new scientific expedition. Taken in concert, they constitute an intriguing challenge to combine modern experimental techniques with appropriate theoretical extensions of quantum mechanics in a resumption of the search for comprehension of this oldest of epistemological issues.

7

MODERN MAN/ MODERN MACHINE

As contemporary information, energy, and materials processing technologies comprising much of modern engineering practice press toward ever more sensitive components and interactive systems, it becomes necessary to protect against extremely subtle external influences and internal interferences. Much microelectronic equipment is now routinely shielded from minute thermal, seismic, electromagnetic, and optical disturbances, and even from background cosmic radiation. Facilities for processing delicate biological substances, rare physical materials, or radioactive ingredients may require exceptional environmental controls and discipline of personnel. In many areas of basic research, from the subnuclear to the cosmological, highly sophisticated pattern recognition techniques must be deployed to extract minuscule signals from background noise patterns. In any of these situations, human operators can impose various individual and collective limitations on the integrity of the devices or data, via carelessness of manipulation or observation, inadvertent thermal, acoustical, chemical, or electromagnetic contamination, or yet more subtle means.

"Any sufficiently advanced technology is indistinguishable from magic."
CLARKE'S THIRD LAW

Beyond the psychokinesis data acquired in conventional parapsychological experiments, there exists a body of anecdotal lore and common experience purporting that certain persons are preferentially prone to precipitate random malfunctions of technical equipment, that crucial machinery tends to fail more frequently in situations where it is most urgently needed, or that physical devices can take on malevolent or benevolent anthropomorphic characteristics. "Gremlin" effects are legendary in military operations, and some test pilots and astronauts will privately acknowledge inexplicable aberrations in their meticulously calibrated guidance, control, and communication systems.

All of this might very well be rationalized in terms of normal random processes overlaid with selective human memory, overactive imagination, or erroneous judgment. Yet, given the trend toward ultrasensitive equipment and the growing reliance on its functional accuracy, and given the admittedly fuzzy but non-negligible background of research results and impressionistic suspicions about paranormal influences, it could be unwise to dismiss categorically the possibility that human consciousness may be capable of more than passive interaction with its microelectronic aids. Can we be quite confident, for example, of the invulnerability of all modern instrumentation, control, and operational equipment to inadvertent or intentional disturbance associated in any way with the psyches of its human operators, especially in periods of intense emotional stress or intellectual demand? Can we be quite sure that the delicate and sophisticated information-processing devices and strategies functioning in juxtaposed microelectronic and mental machinery proceed totally independently, without any interference or cross talk whatsoever? Or should some direct, systematic study be mounted to assess the possibilities of such interactions?

That study is simple to pose; it is not so simple to implement. If real, the phenomena involved are extraordinarily subtle and elusive. It is not incidental that anecdotal experience on such issues has proven difficult to systematize and correlate, that controlled experimentation has so far yielded only sparse pockets of credible data, and that no viable theoretical models have yet emerged for addressing either body of results. None-

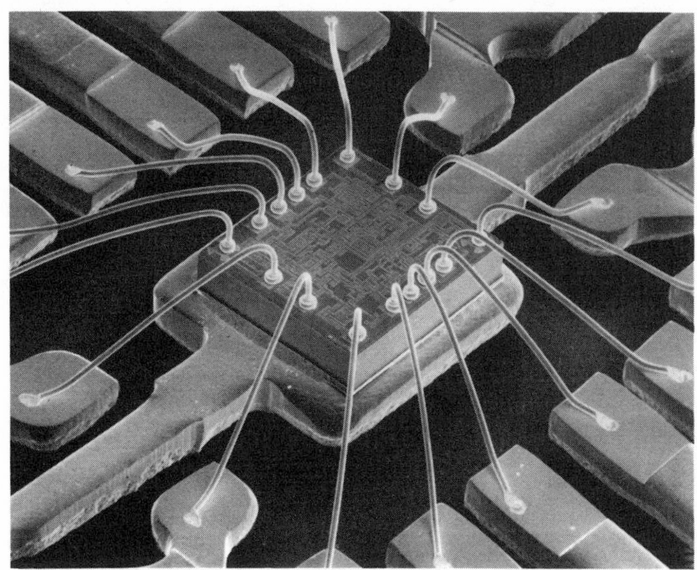

Computer Chip

theless, there is now some hope that the same sharp increase in the sensitivity and sophistication of equipment that has prompted the concerns about its vulnerability to such effects may also hold some promise for their resolution. In particular, the capacity of modern microelectronic circuitry for operation at very sensitive signal levels and for very rapid data acquisition and processing may permit experimental access to domains of interaction where the phenomena of interest are statistically more replicable, so that systematic study becomes more feasible.

In undertaking experiments in this context, we are to some extent extending a trail of more conventional engineering research and development that has variously been labeled "human factors," "cybernetics," "ergonomics," or "man/machine interactions," and that has run parallel to a similar path of study in industrial psychology. Actually, the most productive of such efforts have been those that have compounded insights from the engineering and psychological perspectives in a truly interdisciplinary fashion. But any attempt to address anomalous aspects of the interaction of human consciousness with physical devices clearly must contend with a range of subjective processes that are poorly understood or cannot even be identified *a priori*. Notwithstanding, if the phenomena derive to any degree from conscious or subconscious processes of the human mind, it is important that the experimental equipment and protocols be designed to enhance, or at least not to inhibit, those obscure processes. Very likely, the difference between a sterile experiment and an effective one of equal rigor will lie as much in the impressionistic aspects of its ambience, protocols, and feedback as in the elegance of its instrumentation.

For these and other reasons, it is important that the planning, implementation, and interpretation of this class of research incorporate the insights, intuitions, and intelligence of the human operators who are asked to function as components of the man/machine systems. Given the intrinsic interdisciplinary character of the subject, it is equally essential that the laboratory personnel possess a broad range of experience and be cognizant of forefront research in relevant disciplines beyond their own. Finally, it is helpful if the institution housing the laboratory displays tolerance and support for the unusual tone and special requirements of such research.

3

STATISTICAL SCIENCE

Once Isaac Newton's legendary apple detached itself from the branch, its trajectory toward the ground (or toward Newton's head, depending on the version) was largely determined by universal gravitational effects, modified by a bit of aerodynamics, and would have been very well predicted by the classical dynamics our hero was about to invent. The precise time of detachment of that particular apple, however, depended on a complex combination of biomechanical processes in the stem, along with numerous external factors such as the prevailing wind, humidity, and temperature, and would have been far more difficult to predict by any simple causal theory. For that purpose, it would have been more reasonable to devise some sort of probabilistic model that dealt in terms of the likelihood of the apple falling in any given period of time, or the fraction of all of the apples on that tree that could be expected to detach themselves over that period.

The distinction being drawn is between *deterministic* science, wherein the behavior of some individual object or process may be computed with considerable precision solely on the basis of its initial conditions and the prevailing physical influences, and *statistical* science, wherein detailed knowledge of the individual behavior of any item in a large system is surrendered in favor of some characteristic distribution of behavior of the system as a whole. Clearly, no real physical situation can be perfectly

represented in either extreme. Precise calculation of an individual apple trajectory is limited by a hierarchy of uncertainties concerning the wind pattern, air composition, apple shape and surface characteristics, solar pressure, gravitational aberrations, and so on, and ultimately one must settle for a reasonable degree of approximation that brackets the precise behavior. Similarly, the probabilistic apple-detachment models also must be qualified by first fixing or eliminating certain parameters before accommodating the others within the statistical representation. For example, we might report that "on September 27th, in the Finger Lakes district, Stayman Winesap apples fall from well-pruned mature trees at an average rate of 0.074 per tree, per hour." Note that even in this elaborate qualification, all manner of local meteorological, topological, and biological variations are subsumed within the statistical package.

In either form of representation, the scale of the variation about the average, or mean, property being computed is an important secondary feature. In the deterministic calculation of the apple trajectory, we would like to know the extent of the small uncertainty or inaccuracy that might be imposed on our calculations by not including the subtler dynamical influences mentioned. Any difference in actual behavior that exceeded this range would then constitute an anomaly. Similarly, in the statistical representation of the rate of detachment, the theory should also predict the breadth of the rate distribution about the mean value that is to be expected, or, in statistical parlance, its "variance" or its "standard deviation." Claims that a particular group, or sample, of observations constituted an anomalously large or small fall rate could then be made in terms of the departure of the average measured value from the predicted mean, scaled by the established variance of the distribution. In many statistical situations, other details of the shape of the distribution might also be important, for example, its asymmetry about the mean (skew), its degree of flatness (kurtosis), or yet more detailed aspects.

It is frequently illuminating to display and examine statistical events in a serial form, either to illustrate the temporal evolution of a process, or to show the sequential compounding of a series of events or observations. So, for example, had Newton sat long enough under his tree, he could have accumulated a timewise history of the rate of apple detachments as the harvest season advanced through the late summer and fall months. This would by no means have been a smooth curve, but would have been embellished by various systematic oscillations reflecting diurnal effects, such as the temperature and light cycles, and by numerous less predictable variations caused by weather fluctuations and other irregular disturbances (cf. Fig. I-1[a]). Had he turned his attention to an adjacent tree, he would have found a somewhat different pattern, albeit displaying similar features. If he had then extended his attention to a large number of neighboring trees constituting a well-defined sample population and averaged his observations, a more generic behavior would have emerged that subsumed all of the individual tree characteristics as well as some of the most localized meteorological variations (Fig. I-1[b]). He might then have chosen artificially to smooth this curve to define an empirical baseline against which he could compare the behavior of individual trees.

In the illustrations shown, Newton's original tree displays an earlier and higher yield pattern than that of its neighbor, or of the baseline mean. This could more explicitly be displayed as a graph of the deviation of this tree's fall rate from the

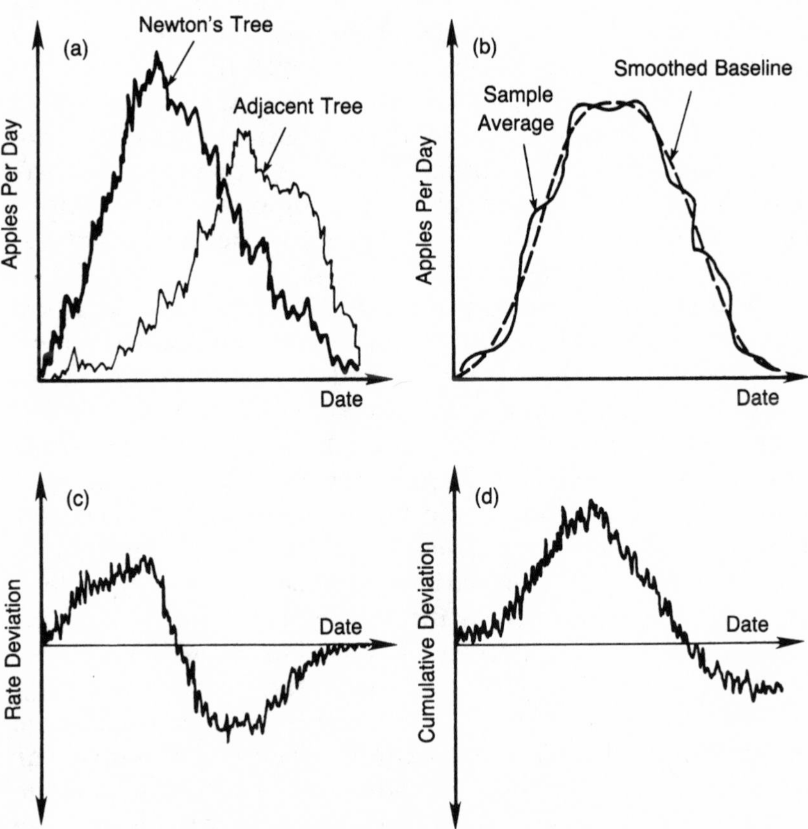

Figs. I-1(a–d) *Temporal Graphs of Apple Detachment Rates*

average behavior of the baseline group (Fig. I-1[c]) or, even better, as a cumulative deviation that tracks the accumulated excess of fallen apples throughout the season (Fig. I-1[d]). Whether this deviation, however represented, is statistically unlikely enough to constitute an anomaly depends on the prevailing variance of all of the individual tree behaviors about the mean, as well as on the scale of the systematic and random oscillations in the individual system. If all of these are relatively small, a deviation of the scale shown could be regarded as an anomaly; if many other trees show departures of similar magnitude, or if the oscillations in the system are comparably high, this deviation would be construed as consistent with the variance of the distribution, and therefore not a significant aberration.

Each of the trees in this hypothetical orchard would, of course, display its own characteristic pattern of deviation from the mean behavior. Some would have higher total yields than the average; some lower. Some would peak their harvest earlier; some later. In any case, each cumulative deviation graph would constitute a sort of "signature" for that tree, displaying its apple-dropping characteristics in an instructive graphical form.

Modern physical, biological, and social sciences have long been accustomed to dealing in such statistical terms. It is impossible to predict the trajectory of an individual molecule in a gas-kinetic swarm, the time of emission of a photon from a particular excited atom in a luminous body, or the lifetime of a given unstable nucleus in a radioactive substance, but the statistical behavior of very large numbers of such events in self-interactive systems can often be predicted with great accuracy. Similarly, the growth patterns or susceptibilities of bacteria, plant cells, or redwood forests, or the reactions of groups of animals or humans to external stimuli must also be represented in statistical terms.

In dealing with such statistical systems, the scientific requirement of replicability of observation needs to be generalized into a somewhat more generous form than that imposed on deterministic systems. In observing individual apple falls, it is reasonable to require that each trajectory conform closely to the predictions of the dynamical theory. In making apple-detachment observations, however, we can only require that the data for the tree as a whole, or for the specified group of trees, conform to the predicted distribution for samples drawn from populations satisfying all of the constraints of the statistical model. No brief succession of individual apple-detachment observations should be expected to replicate one another closely, and any samples drawn outside of the specified group will likely confuse the data. Indeed, it is the ability to provide adequate sample sizes and to limit the number of pertinent parameters that distinguishes good statistical experimentation and establishes a proper basis for statistical replicability.

Any experimentation on consciousness-related phenomena clearly must be conceived, implemented, and interpreted from a statistical perspective. There is no hope whatever of acquiring sufficient *a priori* knowledge of all of the pertinent physical, psychological, and physiological parameters to attempt deterministic empirical experiments or theoretical models. Rather, one must establish certain fixed families of constraints, and then search for average behaviors and characteristic distributions about those averages, within those families. Anomalous effects can then be claimed only in terms of departures of observed means or serial trajectories from theoretical or empirical baseline behavior, scaled by the theoretical variance or by the variance of the baseline data. In most cases, the choices of which parameters to hold fixed and which to explore in any given experiment are numerous, yet clearly critical to the production of intelligible data. The requirement of replicability can be imposed only in the context of effectively delimited experimental situations. The frequent failure of replication that has long plagued systematic parapsychological experimentation and is often cited as evidence for categorical rejection of the claims may rather be indicative of the immense statistical scale and complexity of the phenomena and of inadequately constrained experimental designs.

Given the number and variety of relevant parameters on the human side of the interaction, it would seem prudent to limit any experimental devices or systems employed in consciousness research to those embodying the simplest possible statistical behavior. For all of the reasons noted, apple trees would be a very poor choice for such purposes; their own multitude of temporal and spatial variables would seriously confound the demonstration and interpretation of any putative consciousness-related anomalies, and the establishment of a truly random reference baseline would be very difficult. Far better for the task would be some mechanism based on simple binary statistics, wherein the only available action for each element of the statistical ensemble was some sort of yes/no, on/off, or go/no-go option, with an *a priori* probability of each alternative of precisely 50%, or some other well-established fraction. The anticipated chance behavior of such systems can be computed statistically with great accuracy, so that the significance of any anomalous effect that appeared in a given experiment could be readily assessed. Once again, the scale of departure required for legitimate claim of anomaly would depend on the total scale of the system. If a coin is flipped once, the appearance of a head is certainly not anomalous, since the chance probability of that result for a perfect coin is exactly .5. If it is flipped ten times and all heads are obtained, we begin to suspect that the coin is untrue, or that its fall is being biased by the flipper, since the likelihood of that result occurring by chance is $.5^{10}$, or less than one part in a thousand. If 100 consecutive heads are tossed, the coin becomes highly suspect, or if its integrity is independently confirmed, we face a major anomaly, since the chance probability of that result is less than one in 10^{30}.

For many purposes, a more useful criterion can be posed in terms of the likelihood of reaching some threshold number, for example, of obtaining more than 90 heads out of 100 flips, more than 55 heads, etc. Elementary statistics can predict such composite probabilities by specific binomial formulas or, for very large numbers of flips, by appropriate integration of a generic probability density relation called the Gaussian distribution:

$$f(x) = \frac{1}{\sqrt{2\pi}\,\sigma} e^{-\frac{1}{2}\left\{\frac{x-\mu}{\sigma}\right\}^2}$$

where $f(x)$ is the probability of obtaining x heads and has the form of the bell-shaped curve sketched in Fig. I-2(a). For any given number of total flips, ν, this distribution is completely characterized by its mean value,

$$\mu = \nu/2\,,$$

and its standard deviation,

$$\sigma = \sqrt{\nu/4}\,.$$

Thus, for 100 coin flips, the Gaussian mean is 50 and the standard deviation is 5.

Whether any particular experimental count is to be regarded as anomalous then depends on three factors: how far the count deviates from the chance mean; how large is the standard deviation of the distribution; and what is the criterion for a significant anomaly. For example, if our coin-flipping experiment yields a head count of m, we can first compute the number of standard deviations this departs from the chance mean, μ,

$$z = \frac{m - \mu}{\sigma}$$

and then, by suitable integration of the probability distribution, compute the likelihood of this value of z, or any greater value, appearing by chance (Fig. I-2[b]):

$$p_z = \int_z^\infty f(z)\, dz\,.$$

Such "one-tailed" probabilities are thoroughly tabulated in most statistics manuals. Thus, if our threshold criterion for an anomaly is, say, .05, any z value greater than about 1.65 is "significant." For 100 coin flips, this transcribes to any number of heads greater than 58. If our threshold criterion were .01, we would require 62 heads. If we also chose to regard too *few* heads as an equal anomaly, we would then use a "two-tailed" calculation that computed the chance probability of obtaining either a positive or negative z value of a given absolute magnitude or greater.

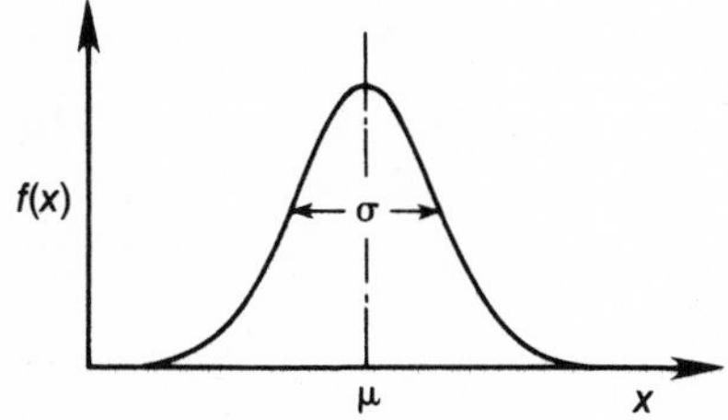

Figs. I-2

Gaussian Distribution

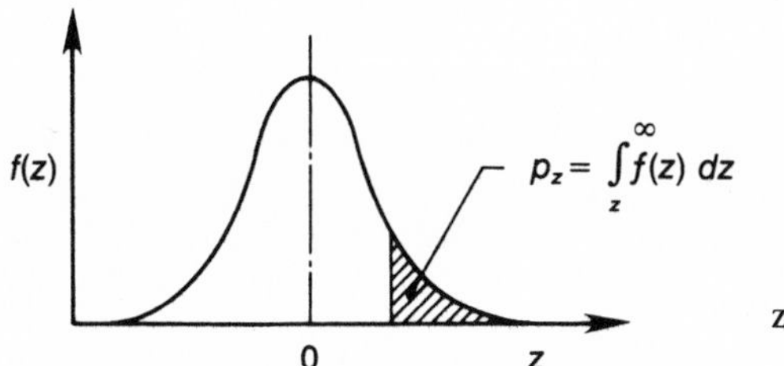

z-Distribution

A more comprehensive assessment of possible anomalies in such binary experiments may be made by comparing a full experimental distribution of counts with that predicted by theory and looking not only for significant shifts of the mean, but for changes in the standard deviation or other variations in the shape of the curve. For example, Fig. I-3 shows comparisons of several hypothetical experimental distributions with the chance Gaussian, each of which might constitute a significant anomaly, but only one of which involves a shift of the mean. Criteria for the significance of the various shape changes illustrated can be developed from more elaborate statistical arguments.

Even if one confines attention only to mean shifts, other forms of data representation may be instructive. For example, the cumulative deviation format used earlier for the apple tree example can be equally illuminating for simple binary systems in displaying the temporal or sequential evolution of the effects. Figure I-4 shows a few examples of such graphs for hypothetical experiments in which there is a) no effect beyond chance; b) a persistent departure from chance expectation; and c) an irreg-

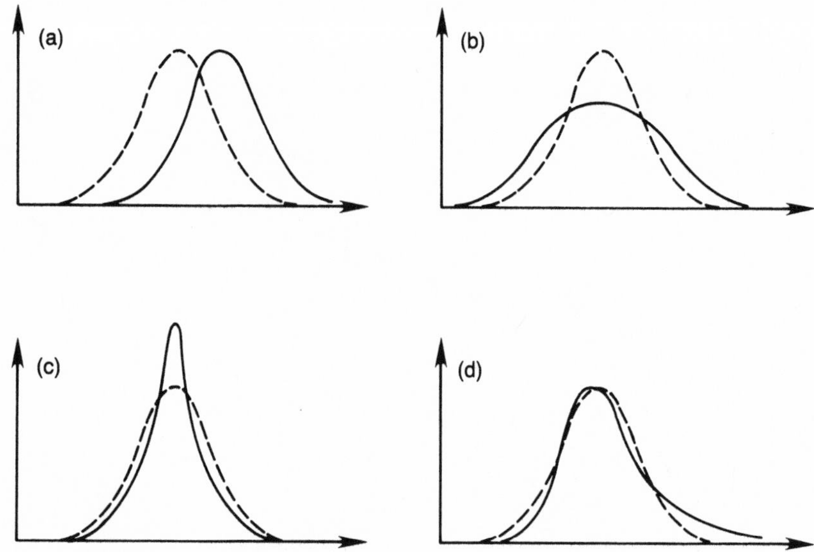

Figs. I-3(a–d)
Experimental Distributions (—) Compared with Chance Gaussians (---)

ular departure from chance expectation. In cases like (b), such displays are particularly effective in emphasizing how relatively marginal departures from chance in small samples of data can compound to increasingly significant effects as the size of the data base grows. We shall return to such representations and their finer interpretations in the context of the experiments described in Sections II and III.

As noted earlier, the basic purpose of any statistical analysis is to derive useful collective properties for systems of interactive elements or processes that are too proliferate, complex, or otherwise unpredictable to compute by direct compounding of the individual element behaviors. For a broad range of physical, chemical, biological, psychological, and sociological systems this is indeed a powerful and productive strategy, but by no means is it omnipotent. For at least as broad a range of highly interactive systems, the individual element behaviors and their interactions with each other are too demanding for even such statistical methods to track, and yet more ambitious and robust modeling techniques are required. Examples of such systems

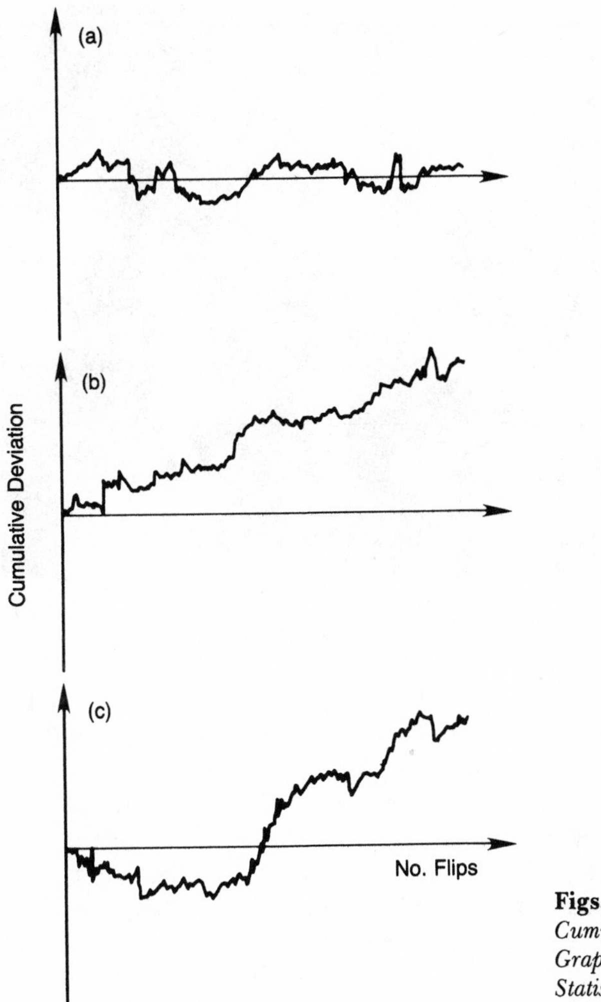

Figs. I-4(a–c)
Cumulative Deviation Graphs for Binary Statistical Systems

include various fluid-dynamical instabilities, such as transitions from laminar to turbulent flow, vortex shedding, or cresting waves; various unstable chemical reactions; or some biological population instabilities. More severe forms appear in various geophysical and meteorological calamities, chemical and nuclear explosions, or medical epidemics. Contemporary theoretical research on such topics as fractals, chaos theory,

Wave Cresting to Chaos

catastrophe theory, nonequilibrium thermodynamics, or nonclassical statistics reflects a growing interest in development of modeling techniques for such eccentric systems.

Potentially interesting for anomalies research are those systems wherein very small disturbances can precipitate disproportionately large changes, thus providing internal mechanisms for major amplification of any consciousness-induced effects. Also of interest would be systems that display exceptional numbers of "outliers" or extreme values of measurable properties; systems that possess self-organizing capabilities; or systems that can both absorb and emit information. Particularly attractive for this study are systems that by their structure produce, process, and represent information in formats comparable to those employed by the human brain, as best we now comprehend it, since these may offer special opportunities for some resonance between themselves and the mental machinery propitious for anomalous interaction of the two. In this category, the recent designs of so-called "neural networks" in computer architecture would seem to be prime candidates.

At whatever level of sophistication the issue is approached, the essential point of this chapter is that any anomaly, by definition, is a significant departure from some pattern of established behavior. Since most established behavior in consciousness-related domains can only be quantitatively represented in statistical terms, and since all statistical formats are themselves constructions of consciousness, it follows that the validity and limitations of the statistical representations must themselves be well defined and well understood before an anomaly can be claimed. It also follows that research attempting to display and comprehend such anomalies should deal with experimental systems that are optimally cooperative for the purpose, either by their statistical simplicity and unequivocality of interpretation or, on quite the other extreme, by their intrinsic vulnerability to small disturbances, their potential for resonance with the human psyche, or both.

9

VECTOR CONVERGENCE

Many other relevant topical vectors could be traced, but those we have sketched in this section bear most directly and influentially on our enterprise:

1) the role of anomalies in the essential scientific dialogue between experiment and theory;
2) millennia of human inclination toward mystical representation and manipulation of reality;
3) the historical stream of scholarly attention to the role of mind and spirit in the physical affairs of the world;
4) the past and contemporary parapsychological research base;
5) the refining role of critical commentary in the qualification of evidence;
6) the conceptual precedents and paradoxes of quantum physics;
7) the rapid escalation of human interaction with "intelligent" microelectronic machinery, and the utility of such devices for modern experimentation;
8) the implications and limitations of statistical science in the representation of experience.

These have so far been traced somewhat separately, as if they proceeded independently toward their intersection on our particular project. Actually, the course of each of them has been, and continues to be, steadily and substantially influenced by each of the others, to the extent that they form a net or web of epistemological support for the enterprise, rather than simply converging upon it.

Historically, all attempts to assess the characteristics and implications of consciousness-related anomalies, like any other issue of broad public interest, have inevitably been conditioned by the prevailing patterns of the cultural environment. Whether addressed in the mystical pragmatism of the Stone Age, the contemplative curiosity of the earliest academics, the magical ambience of the medieval eras, or the resurgent rigor of the Reformation, study of this topic has invariably acknowledged the contemporary credibility criteria and social value system, and this is no less the case in our modern age of technological determinism. To progress toward comprehension of the role of consciousness in physical reality within the current cultural paradigm, it will first be absolutely necessary to establish a body of incontrovertible experimental evidence that is acceptable by current scientific standards. Next, a theoretical model

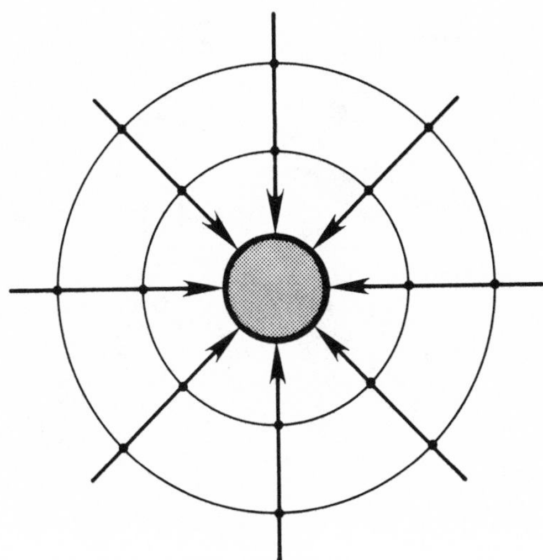

must be proposed that is competent to support a scientific two-step along some rigorous path. Only after both are in hand and functioning harmoniously can consideration of applications and implications, either personal or sociological, credibly be broached. In succession, these are the topics of the following four sections of this book.

In pursuing this scientific high road, however, we must not forget the persistent and ubiquitous relevance of a spectrum of "softer" dimensions—call them mystical, aesthetic, intuitive, religious, spiritual, or whatever—that have permeated the hard analytical thought and rigorous experimentation of all creative science through its many ages, not excluding our own. In some

"Amphitheatre of Eternal Wisdom" (17th Century)

cases this component has been explicitly acknowledged and utilized, in others it has been more subtle and implicit, but its overarching role in creative scientific thought cannot be denied. As Einstein put it:

> The most beautiful and most profound emotion we can experience is the sensation of the mystical. It is the sower of all true science. He to whom this emotion is a stranger, who can no longer wonder and stand rapt in awe, is as good as dead. To know that what is impenetrable to us really exists, manifesting itself as the highest wisdom and the most radiant beauty which our dull faculties can comprehend only in their most primitive forms—this knowledge, this feeling is at the center of true religiousness.
>
> The cosmic religious experience is the strongest and noblest mainspring of scientific research.[49]

Few of the truly great scientists of this or any other era would ultimately deny the importance of these mystical components in their own creative capabilities. What is much more problematic and controversial is whether modern science has the technical, epistemological, and intuitive capacity to demonstrate, comprehend, and incorporate such elusive factors in any rigorous and systematic fashion within its own analytical paradigm.

SECTION TWO

Man/Machine Margins

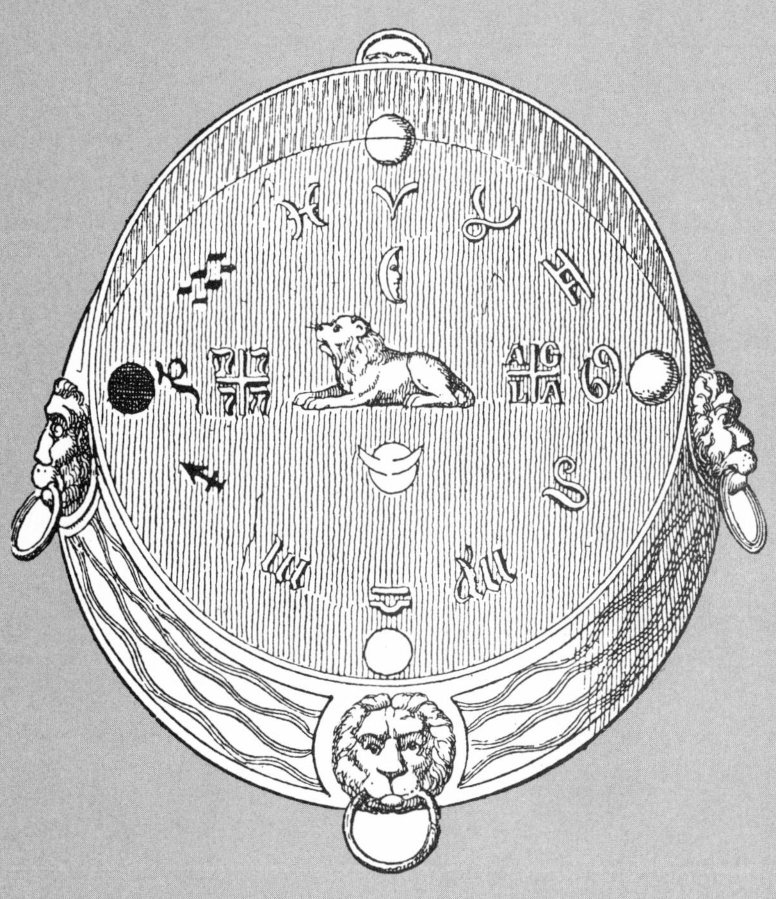

There is no such thing as chance, and what we regard as blind circumstance actually stems from the deepest source of all.

—FRIEDRICH VON SCHILLER, *The Death of Wallenstein*[1]

Wallenstein's Horoscope-Amulet

1

PEAR PROGRAM

The conceptual and strategic vectors outlined in Section I converge on a common need: a program of modern research that directly addresses the role of consciousness in the establishment of physical reality. At their confluence, these diverse perspectives constitute an unresolved chord of enigmatic experience and incompletely substantiated speculation that yearns for definitive evidence and incisive explanation to confirm and clarify its message. Such confirmation may indeed now be scientifically attainable, via clever application of the same modern technology that is potentially most vulnerable to consciousness-related effects. But much more than sophisticated instrumentation will be required. To be credible in our culture, any experiments or theories pursuing this topic must be both scientifically impeccable and pragmatically relevant; yet, to encompass the phenomena at issue, they must also accommodate a variety of intangible and subjective factors that are difficult to address in scientific or pragmatic terms.

The next three sections of this book present an overview and interpretation of one particular response to this challenge, mounted over the past decade in the School of Engineering and Applied Science at Princeton University. The technical and analytical details of this research are covered more fully in a variety of technical reports and journal publications, many of which are referenced at the back. Here we shall only

describe in general terms the principal experiments and modeling methods, along with the ambience and style of the laboratory in which they are deployed.

The Princeton Engineering Anomalies Research (PEAR) program was formally established in the spring of 1979, but the basis for it actually had been laid two years earlier in the form of an undergraduate independent project. The student, an electrical engineering and computer science major, had approached one of the authors (RGJ) with a request for supervision of the design and implementation of an electronic device suitable for study of low-level psychokinetic effects similar to those reported in the recent parapsychological literature.[2] Torn between a personal intellectual skepticism and the potential pedagogical benefit of such a project, the adviser first agreed merely to thorough bibliographical review of the topic and only subsequently to the proposed experiment. Two years later, when this student graduated, the adviser was still far from convinced of the validity of the purported phenomena, but he was now becoming progressively more concerned about the potential vulnerability of certain technological systems to effects suggested by these pilot experiments. To assess such possibilities more systematically, he undertook to organize, fund, and supervise a more substantial research program directly addressing man/machine anomalies.

The historical details of the genesis and evolution of this program, and of its interactions with the academic, industrial, and federal communities, could provide the material for yet another volume, but we cannot afford this excursion here. Let us simply note that the substance and style of this work have severely tested the sacred tenets of freedom of inquiry and humility in the face of experimental evidence that the scientific and academic communities have long espoused. Because of these sensitivities, it has been necessary to impose at least as stringent a discipline on the public posture of the program as on its technical rigor. No impact on the academic curriculum has yet been allowed beyond the occasional accommodation of a few other selected student projects. All funding has been provided by private foundations; no government or university support has been involved. The results of the research have heretofore been reported only in professional journals and technical reports, in oral presentations to professional meetings, and in published proceedings thereof. No media coverage has been authorized in any form. Thus, this book constitutes the first major representation of the work to a more general audience.

As its name implies, the tactical approach of the PEAR program has been to apply the techniques of modern engineering science to systematic study of selected aspects of consciousness-related anomalies that appear pertinent to contemporary science and engineering practice and amenable to controlled laboratory study. Consistent with this purpose, and in contrast to most contemporary parapsychological research, the experiments have focused primarily on physical and technical parameters of the devices, processes, and protocols employed. Consideration of psychological or physiological correlates has been subordinate to the accumulation of very large data bases provided by a relatively small number of regular participants or "operators." Over all, some fifty men and women of various ages and backgrounds have contributed to the data base, including the authors and other members of the laboratory staff. None of these has claimed extraordinary or "psychic" abilities. It has been our policy to maintain the anonymity of all operators, and to provide no compensation beyond travel expenses and their own satisfaction in participating in the work. No screening, training, or psychological testing has been employed, but every effort has been made to provide a comfortable and relaxed environment, to include the operators in technical discussions, and otherwise to regard them as full collaborators in the research.

In its present configuration, the program consists of three distinct but synergistic sectors of study. The first, which is the subject of this section, is an ensemble of experiments investigating operator-related man/machine anomalies—interactions of human consciousness with various physical devices, systems, or processes resulting in statistical output characteristics significantly deviant from those expected on the basis of known scientific mechanisms. The second, described in Section III, concerns precognitive remote perception—the acquisition of information about locations remote in distance and time and inaccessible by any known sensory communication channel. The third segment, outlined in Section IV, is the development of a theoretical model useful for correlating both classes of experimental data, for designing more incisive experiments, and for explicating the general phenomena on fundamental grounds.

2

BINARY BIAS

The requisite characteristics for the first category of experiments—the interaction of human operators with potentially vulnerable physical mechanisms—derive from various aspects of the empirical experience and statistical logic sketched earlier:

1) The device or system with which the human operator attempts to interact should embody some random physical process that is transcribed into a readily discernible output distribution of data in a form amenable both to direct quantitative recording and to attractive feedback for the operator.

2) The statistical character of this output distribution should be simple, direct and, ideally, theoretically calculable.

3) The experimental equipment should permit very high signal-to-noise discrimination and provide extreme protection against technical malfunctions, environmental artifacts, or other spurious influences.

4) The entire experimental system should be capable of rapid acquisition of very large data bases.

5) Extensive and regular calibrations should confirm the stability of the system and its conformity to theoretical expectations.

6) Experimental protocols should optimize operator incentive and ease of operation, emphasize the primary parameters of interest, and provide further discrimination against spurious results.

7) The data collection, storage, and computation facilities should provide redundancy of records and be capable of rapid and precise rendering of experimental data into the most instructive quantitative representations of results.

8) Beyond its technical sophistication, the physical and social ambience of the experimental facility should be conducive to uninhibited participation of the operators in both the performance and interpretation of the experiments.

In any practical implementation, various optimizations and trade-offs among these desirable characteristics must inevitably be faced, but to be effective and credible in this type of experimentation, the composite system must retain substantial strength in each of these categories.

Of all of the statistical processes that might underlie experiments of this class, the binary combinatorials noted in Section I-8 are the simplest, both conceptually and computationally. For example, in a rudimentary experiment, an operator might flip a coin many times while willing an excess of heads, and the distribution of results might be compared with the theoretical Gaussian prediction. One could then calculate by statistical argument the likelihood that the achieved distribution could have occurred by chance or, conversely, the likelihood that some volitional bias had been introduced into the basic binary event. Such an experiment would be seriously flawed, however, in at least four respects: a) no real coin would be perfectly balanced; b) the operator's style of flipping would inevitably be nonuniform; c) random environmental disturbances could affect the process; and d) the rate of data acquisition would be tediously slow.

All of these limitations can be greatly relieved by employing microelectronic circuitry to perform the essential functions of "flipping," counting, displaying, and processing. One class of device that has become rather common for such experiments is called a Random Event Generator (REG). Typically, such machines are based on a source of electronic "white noise" generated by some random microscopic physical process, such as a thermal electron current in a solid-state diode, a gaseous discharge, or a radioactive decay. Electronic logic circuitry then transforms this noise into a regularly spaced string of randomly alternating binary pulses, or "bits" (+, −, +, −, −, +, −, +, +, +, −, −, . . .), that subsequent circuitry then counts, displays, and records. All of this can be performed with great precision and safeguard against artifact, at very rapid rates, constituting a much more serviceable electronic "coin-flipping" device toward which the operator may direct his intention.*

Our particular REG apparatus, which is a refinement of the original student project, utilizes as its random source a commercial microelectronic noise diode unit commonly incorporated in a variety of communications, control, and data-processing equipment. In addition to the filtering, sampling, counting, and display circuitry that renders this noise into an output distribution of binary counts, the system entails extensive fail-safe and calibration components that guarantee its integrity against technical malfunctions and environmental disturbances. A photograph of the device as it appears to the operator, along with a block diagram of its logic circuitry, is shown in Fig. II-1. Full details of the design of this machine are available on request.

The REG constructs its binary output string by sampling the electronic noise pattern at preset regular intervals. If the noise signal happens to be positive (greater than its mean value) at the time of sampling, a positive pulse or bit is produced; if the noise signal is negative at that time, a negative bit is produced. The basic experimental data group, called a *trial*, comprises some given number of these electronic bits. By dial setting, the machine may be instructed to form 20, 100, 200, 1000, or 2000 bits for each trial. Similarly, it may be told how rapidly to take

* Throughout this text, generic pronouns refer equivalently to any of our operators, male or female. This implies no hierarchy of numbers, effectiveness, or contribution to the program. To the contrary, the mix of perspectives and styles among our participants and laboratory staff has been an important factor in the design, operation, and interpretation of the experiments, and is implicitly acknowledged in the formulation of the theoretical model.

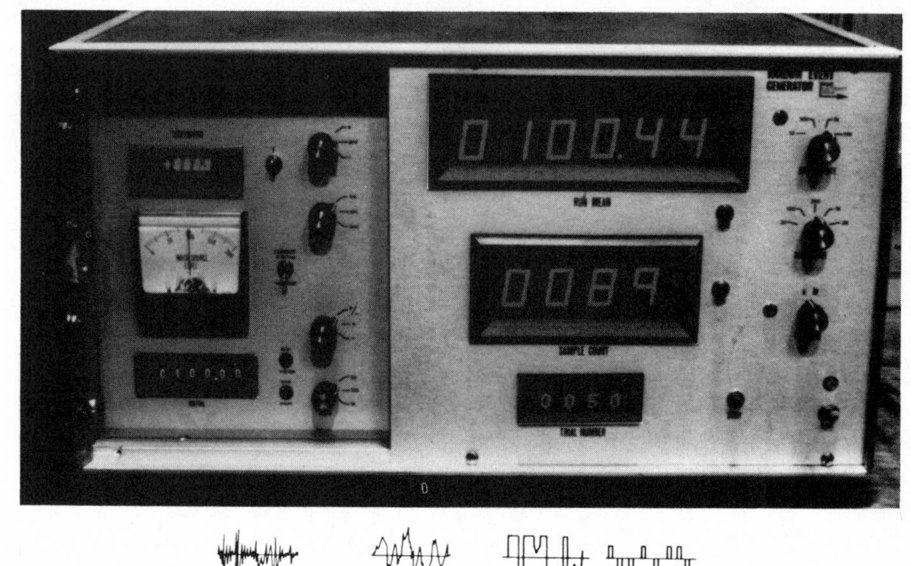

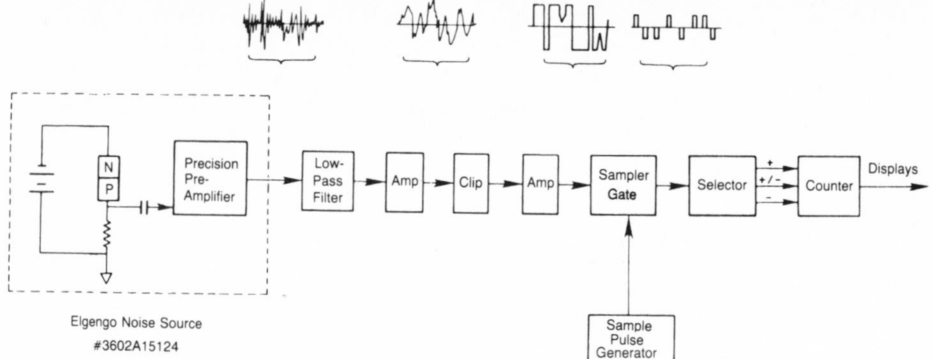

Fig. II-1 *Microelectronic Random Event Generator*

the samples—10, 100, 1000, or 10,000 times per second—and whether to count only positive bits, only negative bits, or to count those bits conforming to a regular alternation (+, −, +, −, +, −, . . .). This last, alternating mode serves to eliminate any effects of residual bias in the basic noise pattern and is employed for all data reported below.

For any combination of these options, the individual trial scores immediately appear as illuminated digits on the face of the REG machine and are simultaneously printed on a continuous paper tape and transmitted to a computer file where they are stored for subsequent processing. Also displayed are the accumulated mean of all trials so far generated, and the current trial number. These three digital displays serve as the main feedback for the operators, informing them of their progress in the intended task, much as in a typical biofeedback display.

Extensive and frequent calibration procedures confirm that unattended operation of this device provides output distributions of trial scores that are in excellent agreement with the statistical predictions. Figure II-2 displays a frequency-of-count distribution of some 55,000 200-bit calibration trials superimposed on the appropriate theoretical Gaussian curve. The mean of the calibration distribution, its standard deviation, tail-populations, and various other subtler features are all well within the statistical variations to be expected by chance for this size data base. A variety of yet more sophisticated statistical tests and electronic calibrations have further confirmed the integrity of the apparatus for the purpose designed.

Like the rudimentary coin-flipping experiment, the primary goal of the REG studies is to explore correlations of statistical biases in the output count distributions with prestated intentions of the operators. To establish any such biases in the most incontrovertible terms, the experimental protocol must allow the accumulation of large blocks of data wherein all controllable

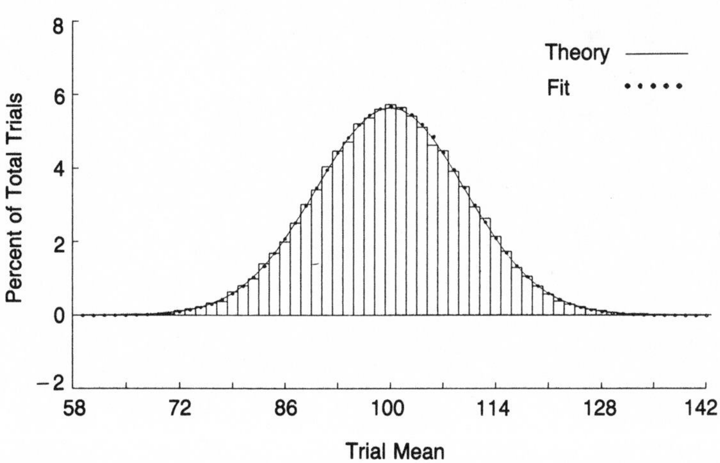

Fig. II-2 *Count Distribution from 55,000 REG Calibration Trials*

secondary parameters are held fixed, and this primary variable—the intention of the operator—is explored in some manner that selectively factors out any unforeseen or uncontrollable spurious influences. The strategy we employ to accomplish this is termed a *tripolar protocol,* in which data taken under an intention of the operator to achieve higher than chance trial counts, henceforth labeled PK$^+$; data taken under intention to achieve lower than chance counts, PK$^-$; and baseline data taken with no conscious intention, BL, are interspersed in some reasonable recipe, with all other operating conditions, including the physical position of the operator, remaining the same. Only if systematic deviations of major statistical magnitude are observed among the ensuing three streams of data are correlations of the results with operator intention claimed. Thus any artifact of the equipment or environment cannot affect the results over large blocks of data unless it is itself somehow correlated with operator intention.

One important strategic option in this tripolar protocol is the number of trials that are blocked together during acquisition and processing of the data. In essence, these block sizes are set by a compromise between the establishment of clear systematic trends in the data, and the stamina and attention span of the operators. Largely on the basis of empirical experience with both of these factors, three levels of data agglomeration are maintained. The smallest of these is called a *run*—the number of trials strung together in a single effort of the operator, without pause for recording or other break in attention. The bulk of our REG data has been acquired in runs of 50 trials, although a substantial base of data has also been generated in 1000-trial runs. Once the run length is selected, it remains constant throughout the experiment. A second, more flexible grouping of trials is termed a *session* and consists of the number of runs the operator completes in any one period of access to the machine. The choice of session length is left largely to the preference of the operator, although certain minimum block requirements are imposed, depending on the prevailing secondary parameters, to preclude optional stopping effects from distorting the data.

The largest and most important unit of data concatenation is the *series,* which typically consists of 2500 or 5000 trials in blocks of 50 or 100 runs, or 3000 trials in blocks of 3 runs, under each of the three intentions, PK^+, PK^-, and BL. The series length is also fixed at the start of an experiment. Operators usually take two to six weeks to complete a full tripolar series, although some have taken as little as one day or as long as a year. Despite the major commitment required of an operator to generate such a series, we have found that this is the minimum base of data from which consequential systematic trends can reliably be extracted from the inherent statistical variations, and therefore is the unit on which we perform most of our statistical assessment.

Beyond the primary variable of operator intention, a number of secondary parameters suggest themselves for systematic experimental survey. For example, the various technical options provided by the machine for sampling rates, sample sizes, or counting of only positive or only negative bits could conceivably have psychological implications for the operator's performance of the primary task, or even more fundamental relevance to the nature of the anomalous interaction itself, and thus merit some exploration. Possibly even more instructive could be differences in the performance of the operator when allowed to choose the direction of intention and when assigned the direction by some other person or random process. To explore this sensitivity, two versions of the tripolar protocol are employed. In the *volitional* mode, the operator decides in advance the sequences for interspersing his PK^+, PK^-, and BL runs, subject only to a constraint of a minimum of five consecutive runs in each condition. These intentions are, of course, prerecorded in the data base. In the *instructed* mode, the operator activates an ancillary random criterion in the REG before each run that specifies in advance the direction he should attempt and automatically records the assignment. (Thus, in this mode, the number of runs performed under each intention in a given series may not be exactly balanced.)

A similar division of protocol has also been established to explore the importance of the operator controlling the time of initiation of each trial. In the *manual* mode, the operator initiates each trial at his discretion; in the *automatic* mode, he initiates the first trial at his discretion, after which the remainder of the trials in the run follow automatically in a regularly spaced sequence.

The following table summarizes the hierarchy of experimental options and protocols available for these microelectronic REG experiments:

TABLE II-1

REG Experimental Options

Experimental Units

Trial:	Number of Electronic Bits in Basic Sample (20, 100, 200, 1,000, 2,000)
Run:	Number of Trials in Single Effort of Operator (50, 100, 1,000)
Session:	Number of Trials or Runs in One Period of Experimental Effort (Operator Preference)
Series:	Minimum Data Blocks for Statistical Processing (2,500, 3,000, 5,000 Trials per Intention)

Primary Variable:

Operator Intention (PK^+, PK^-, BL)

Secondary Parameters:

Instructed/Volitional Assignment of Intention
Automatic/Manual Sequencing of Trials
Bits per Trial (20, 100, 200, 1,000, 2,000)
Sampling Rate (10, 100, 1,000, 10,000 Bits per Second)
Counting Mode (+, −, +/−)

In an actual experiment, the operator first decides all of the options listed in the table, records his choices in the logbook and on the computer, and sets the REG controls and computer indices accordingly. If the experiment is to be volitional, he then decides his direction of effort and records that; if instructed, he activates the random instruction program and acknowledges its result. Then, seated a few feet from the machine, he initiates its operation by the remote switch and endeavors to influence its output to conform to his stated intention, by whatever strategy he feels effective. Baseline runs are interspersed in some reasonable fashion in appropriate numbers to balance the PK trials.

The experiment proceeds within a homey decor of panelled walls, carpeting, and comfortable furniture, with options for variable lighting, background music, and snacks. The laboratory staff maintain a minimal presence during the actual operation, but openly discuss the experimental goals, gear, and protocols prior or subsequent to the session in as much detail as requested. They display cheerful, albeit professional attitudes of openness to any results or lack thereof, and encourage the operators to take a playful approach to the task, rather than being overly concerned about the outcome. The computer software for specifying, initiating, and recording experimental runs is designed to be "user friendly" and to encumber the operator as little as possible with analytical tasks.

Operators are encouraged to engage in the experiments whenever they are in the mood, or when they wish to explore correlation of their performance with particular attitudes or emotional states. As mentioned, the length of a session is left to operator preference and varies considerably. Using the automatic mode at a counting rate of 1000 bits per second, most operators can complete a 500-trial tripolar session in less than one hour. Experiments using slower counting rates, larger sample sizes, longer run lengths, or manual operation are correspondingly slower.

Quantitative evaluation of experimental data generated in this fashion must be consistent with the qualitative nature of the results. The most reliable indicators of the statistical character of the data are the distributions of trial scores obtained over full experimental series. If these distributions closely resemble those of the calibrations and theoretical calculations, then quite simple statistical indicators may be applied in their analysis. However, if major distortions in the distribution shape are observed, more sophisticated treatment would be required. Figures II-3(a,b) illustrate actual trial score distributions for a sequence of 5000 BL, PK^+, and PK^- trials of one particular operator, superimposed on the theoretical Gaussian curves. In each case, the Gaussian character of the distribution has been preserved, with the variance and higher moments statistically indistinguishable from the theoretical values. However, while the mean of the baseline data conforms well to chance expectation, the PK data show small but perceptible shifts of the means in the intended directions with respect to the theoretical prediction. Figure II-3(c) shows analytical Gaussian fits to these same PK^+ and PK^- data, better displaying the scale of these anomalous mean shifts. Small as these may appear to be, for the size of this data base their likelihood of occurring by chance is extremely small.

Since the calibration data conform so precisely to theoretical Gaussian predictions, and since the experimental distributions for the PK^+, PK^-, and BL intentions also remain well fit by slightly displaced Gaussians, very simple statistical measures may be employed to assess the significance of the observed shifts in the means of these trial score distributions. This is fortunate, for given the implications of the phenomena involved, it is important that any systematic anomalous behavior clearly emerge from the intrinsic random noise background in a sufficiently robust fashion that it does not depend on subtle statistical argumentation. For example, we may now utilize the "*z*-score" criterion mentioned in Section I-8, computing the number of theoretical standard deviations, σ, by which the mean of any given experimental trial score distribution departs from the theoretical expectation for the mean, μ, suitably modified by the square root of the number of trials constituting the distribution, n:

$$z = \sqrt{n}\,\frac{m - \mu}{\sigma}\,.$$

The statistical significance of this mean shift can then be extracted from tabulated calculations of the appropriate probability integrals. For the particular data shown in Figs. II-3(a,b,c), the mean trial scores for PK^+, PK^-, and BL, their corresponding z values, and their probabilities of chance occurrence, p_z, are listed in Table II-2, along with the actual number of trials yielding counts in the intended directions. In

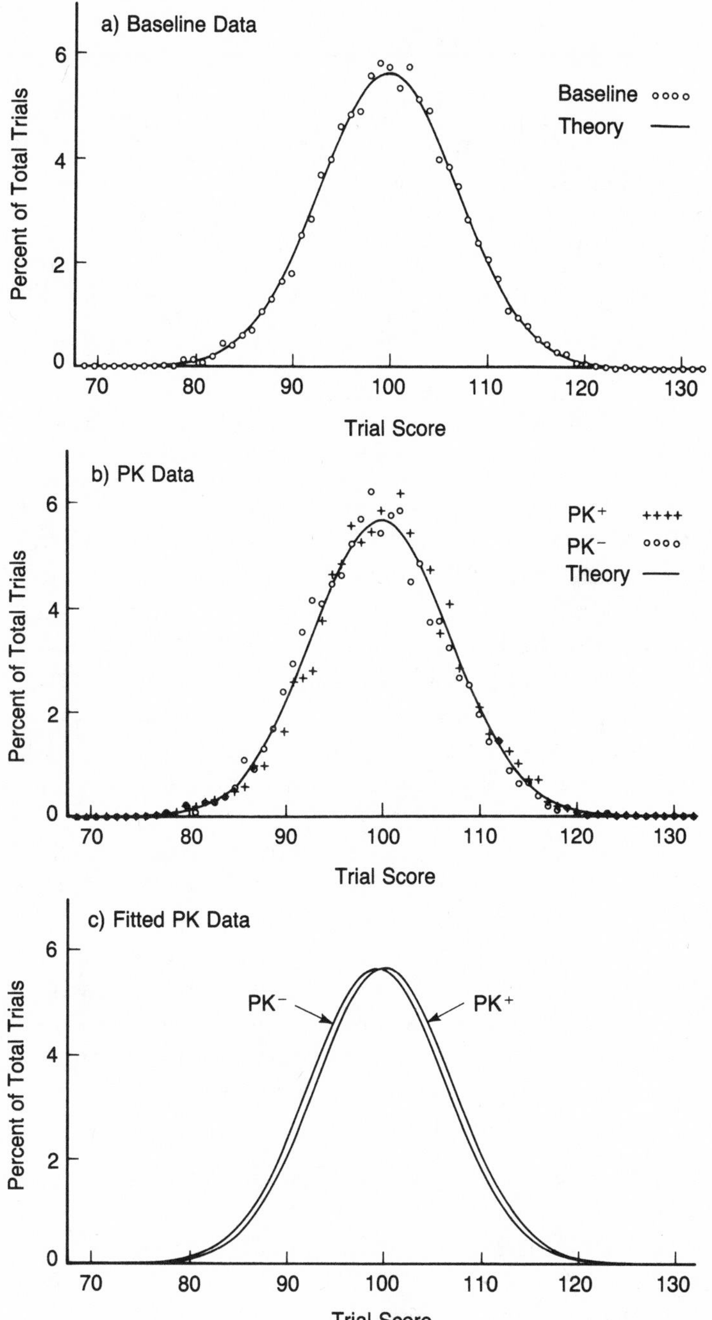

Figs. II-3(a–c) *REG Count Distributions: Operator 10, First 5000 Trials, BL, PK^+, PK^-*

TABLE II-2

REG Data of Fig. II-3

Operator 10: First 5000 BL, PK$^+$, PK$^-$

Intention	No. of Trials	Mean	Std. Dev.	*z*-Score	Prob.	+/− ratio*
BL	5,000	99.898	7.053	−1.020	.154	2327/2386
PK$^+$	5,000	100.279	7.043	2.790	.003	2453/2257
PK$^-$	5,000	99.571	7.036	−4.290	9×10^{-6}	2231/2500
ΔPK	10,000			5.006	3×10^{-7}	

* Number of trials with counts greater than chance mean/number less than chance mean.

this case, the overall means of the PK$^+$ and PK$^-$ data differ from the theoretical mean to extents that far exceed reasonable chance expectation for a data base of this size, while the data acquired under the null or baseline intention, like the unattended calibration data, yield a mean well within chance expectation. Of particular interest is the overall separation of the means of the PK$^+$ and PK$^-$ data from each other (ΔPK), which in this case is unlikely by chance by less than one part per million.

It thus appears that in these particular experimental data the operator has altered the .50/.50 probability of occurrence of the basic binary units by a very small but nonetheless very significant amount, in accordance with his prestated intentions. Specifically, the PK$^+$ bias is of the order of .501/.499, which would be equivalent to inversion of roughly two bits per thousand from chance expectation over the very large number of bits processed. The PK$^-$ bias is about twice as large.

A hierarchy of questions then follows:

1) With what consistency can this operator repeat the accomplishment?

2) How uniformly is the bias sustained throughout a given experimental series?

3) Can other operators achieve comparable results?

4) To what extent do the results depend on the various operational parameters?

5) To what extent do the results depend on the particular noise source or logic circuitry?

6) How do combined efforts of more than one operator reinforce or detract from one another?

7) To what extent do the results depend on the physical separation of the operator from the machine?

8) To what extent do the results depend on psychological or physiological characteristics of the operator?

Such questions have largely defined the course of our experimental program in this area. Some have been pursued to the extent that answers can now be given with some authority of data; others are under active study; yet others have not yet been formally addressed.

3

SIGNATURES AND SENSITIVITIES

Another useful form of representation of this type of data is the "cumulative deviation" graph described in Section I-8, that compounds in serial fashion the accumulated excess or deficiency of binary counts compared to theoretical expectation as the total number of trials increases. Figure II-4 shows such a graph for the same PK^+, PK^-, and BL data displayed in Figs. II-3(a–c) and Table II-2. In this format, all three of the experimental traces display the stochastic variations to be expected in this sort of random process, but whereas the baseline trace meanders close to the theoretical expectation,

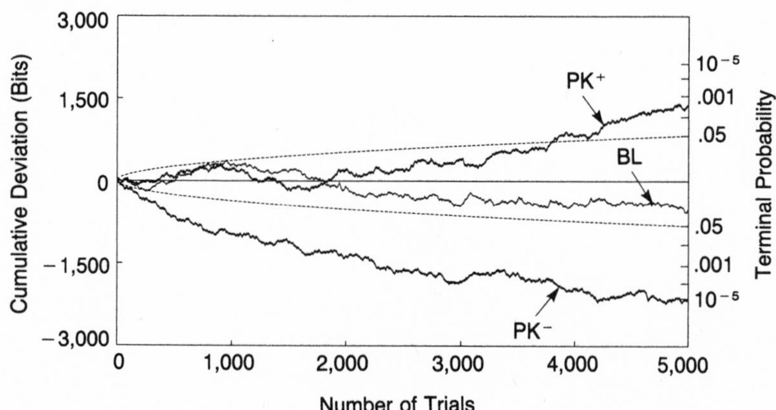

Fig. II-4 *REG Cumulative Deviations: Operator 10, First 5000 Trials, BL, PK^+, PK^-*

the PK⁺ and PK⁻ data accumulate growing systematic deviations from the chance mean as the trials aggregate. In this figure, the dashed lines plot the 5% chance criteria as a function of the number of trials, while the scale on the right border indicates a more complete range of the terminal chance probabilities. Consistent with Table II-2, the PK⁺ and PK⁻ data reach cumulative separations that would have occurred by chance less than once in a million similar experiments.

Also displayed by this cumulative deviation format is the striking uniformity of performance of this operator over the course of these trials. Clearly, the overall results have been achieved not by any short successions of extraordinarily large anomalies, but by relatively steady small biases of output in both the PK⁺ and PK⁻ efforts that can be well approximated by simple linear slopes over most of their range. Note also the slight imbalance of achievement in favor of the PK⁻ efforts, leading to a substantially lower probability of that terminal score.

Our question regarding replicability of such accomplishment can now be addressed by concatenating such cumulative deviation graphs for a succession of experimental series performed by this operator over extended periods of time. Figure II-5 displays this operator's data for all fifteen comparable experi-

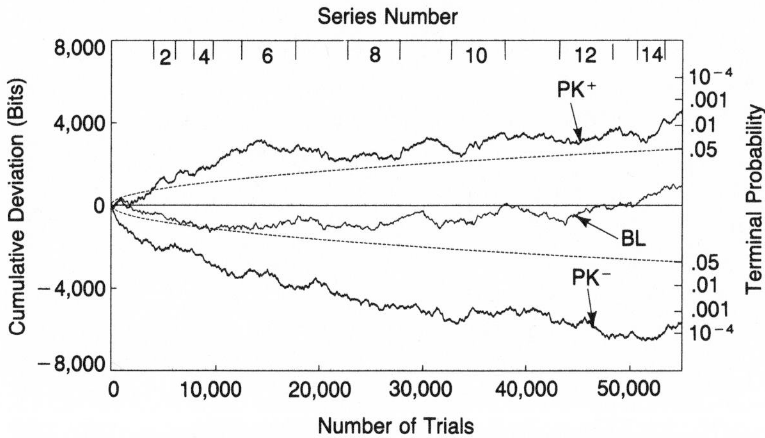

Fig. II-5 *REG Cumulative Deviations: Operator 10, All Data*

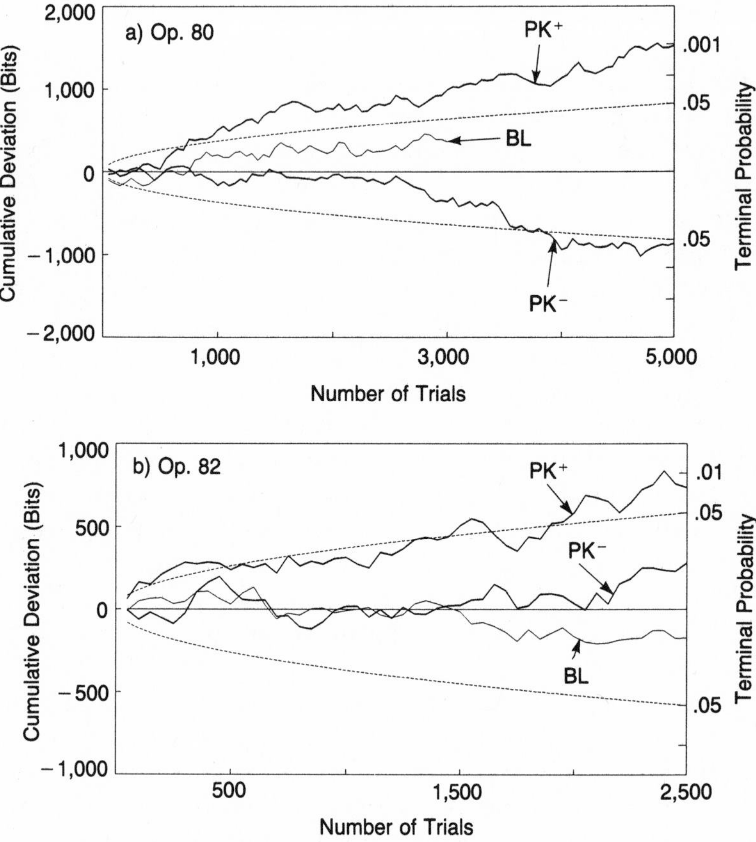

Figs. II-6(a–d) *REG Cumulative Deviation Signatures: Four Operators*

mental series performed over a six-year interval. Within the intrinsic statistical noise of these data, it is clear that the overall character of this operator's achievement generally persists from series to series, as well as within each series, despite the inevitable span of psychological and environmental conditions that must have been subsumed over so long a time and so immense a data base.

It is, of course, imperative to ascertain whether other operators can achieve comparable effects and, if so, how consistently and with what individual character. At the time of this writing, 47 different operators have generated 290 complete REG experimental series, totalling over 2.5 million trials, or a total of roughly 670 million bits processed. In general, it is found

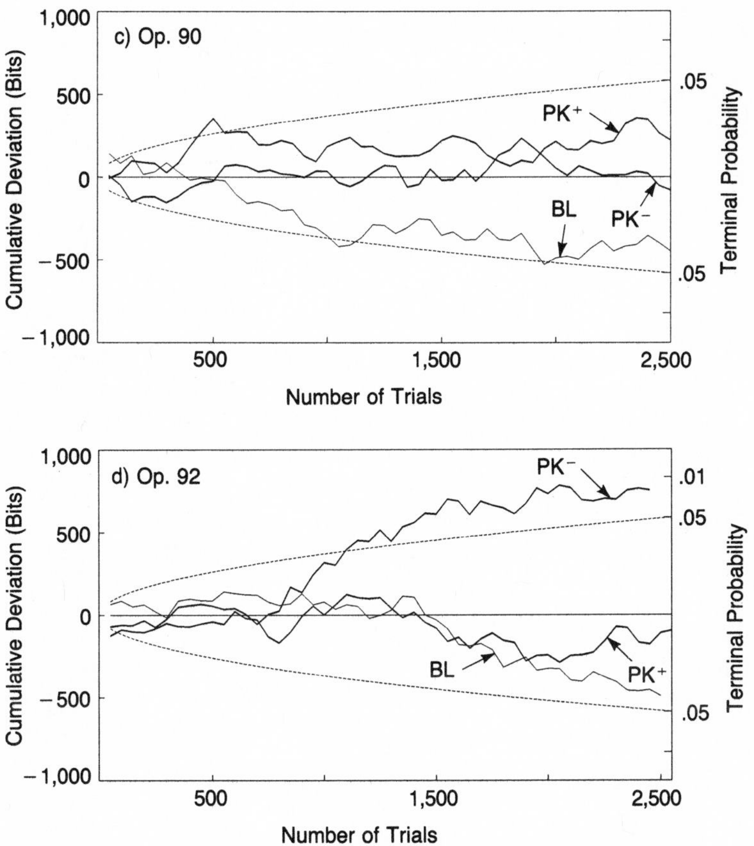

that most operators' cumulative deviation graphs take characteristically different forms that are relatively self-consistent over many consecutive series. For this reason, we have come to refer to such graphs as "signatures" of these operators. A few examples of operator signatures are shown in Figs. II-6(a–d). As can be seen, some operators, as in our first example, show correlations of results in both directions of intention. Others achieve in only one direction, while the reverse effort remains within chance expectation. Yet others achieve in neither direction. Some operators bias the output in the direction opposite to their intention in one or both of their PK efforts. The degree of replicability that can be achieved is again illustrated in Figs. II-7(a–d), which presents the results of three separate series

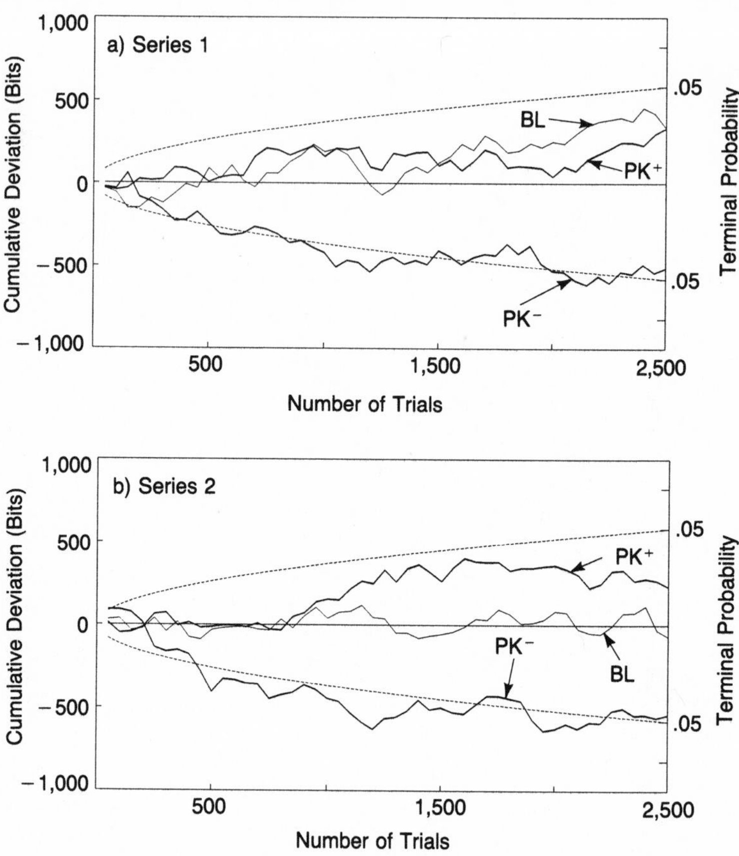

Figs. II-7(a–d) *REG Signatures: Operator 16, Three Series and Combined Data*

by one operator, generated over a two-year period, along with a graph of the full data base.

For simplicity of presentation, our commentary will refer specifically to the largest subset of data for which the technical parameters of sample size and run length have been held fixed—at 200 bits per trial and 50 trials per run, respectively. This encompasses efforts by 33 operators completing 87 series, totalling over 250,000 trials per intention, or approximately 155 million bits. A summary of the results of this body of data is presented in Table A of the Appendix. Before attempting to interpret these, it should be recognized that for any arbitrary statistical significance criterion, a certain number of results exceeding that criterion are to be expected by chance. For

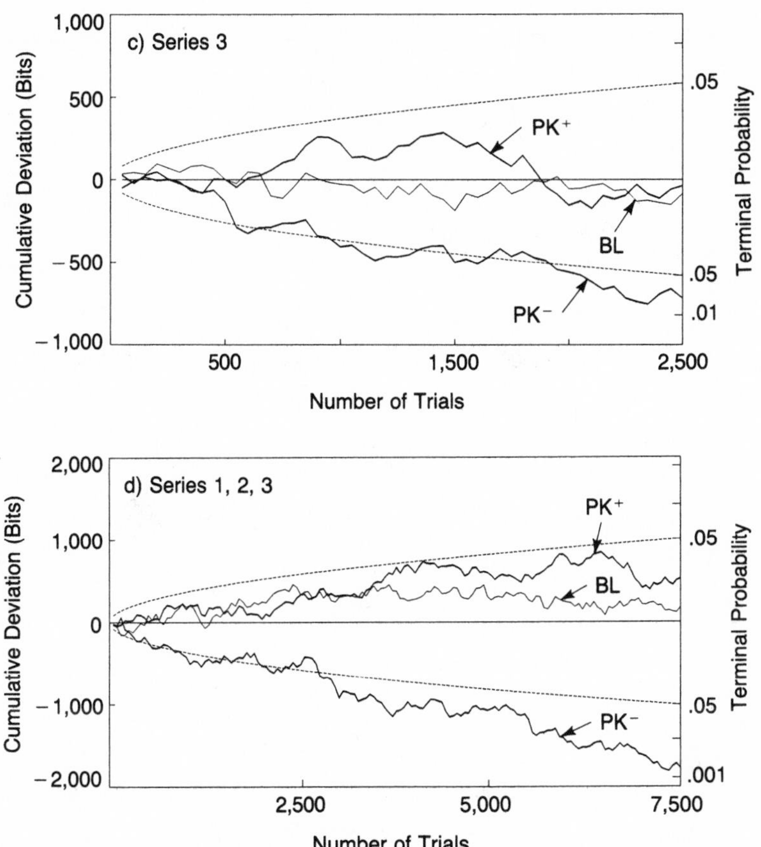

example, 5% of the 87 series, or roughly 4, would be expected to exceed the .05 criterion for each direction of intention, or to display a similar difference between the PK^+ and PK^- results. In point of fact, this family of operators achieved 10 significant series for PK^+, 8 for PK^-, and zero for BL. By the ΔPK criterion, 10 series exceeded chance expectation. Closer inspection of the individual operator data bases indicates that 3 of the 33 achieved significant overall PK^+ performance, 3 achieved significant PK^- performance, and 5 achieved significant ΔPK, compared to the 1 or 2 expected by chance in each case. By a somewhat different criterion, namely the number of series showing results that correlate with intention to any degree, 50%, or 44.5 would be expected by chance in each

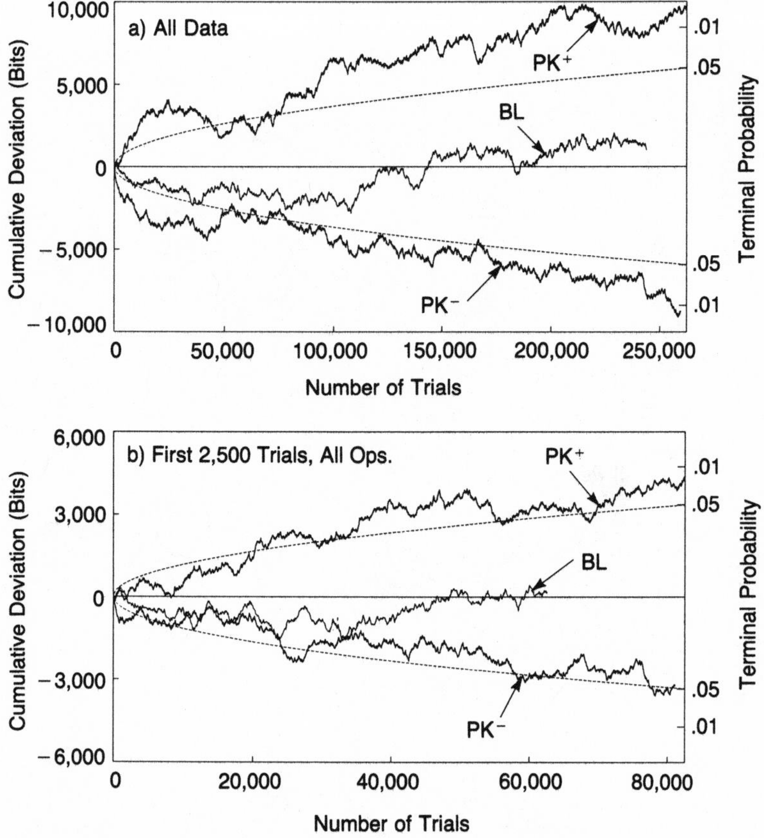

Figs. II-8(a,b) *REG Grand Cumulative Deviations: All Operators*

direction of effort, or in ΔPK. As shown in Table A, 53 of the PK^+ series, or 61%; 56 of PK^- series, or 64%; and 56, or 64% of the ΔPK separations were consistent with the operators' intentions. All of these results thus constitute statistical anomalies that merit more detailed exploration. (More complete tabular and graphical data on all individual operator performances and more thorough statistical analyses are available in a technical report.[3])

Despite the wide disparity in operator achievement, including the fact that a majority of the group does not achieve individually significant results, combination of their data into a grand cumulative deviation graph displays instructive characteristics in its own right (Fig. II-8[a]). Despite the occasional reversals and flat regions imposed by the less successful efforts, the overall mean shifts for both PK^+ and PK^- intentions nevertheless accumulate to significant departures from chance expectations, while the baseline remains close to the theoretical mean. Figure II-8(b) shows a similar combined graph for all of these operators, using only the first 2500 tripolar trials of each to provide an equally weighted combination of their individual efforts.

Given these indications of operator-specific cumulative deviation signatures, the next question in the tactical hierarchy concerns the sensitivity of individual operator performance to the various operational parameters of the experiment, such as the volitional/instructed and automatic/manual options. It might have been hoped, both for parsimony of the experimental matrix and simplicity of the theoretical modeling, that such secondary variables would not bear heavily on the results, or at least that their influence would be generic. Unfortunately, such appears not to be the case; rather, the importance of such aspects seems to vary greatly from operator to operator.

Regarding sensitivity to the volitional/instructed option, for example, Figs. II-9(a,b) compare the signature of one particular operator working solely in the instructed mode, where the direction of intention is assigned by a random process, with a comparable body of data taken in the volitional mode, where he is allowed to choose the intended direction for each block of five runs. In the instructed condition, this operator shows a consistent "psi-miss" pattern, inverting the performance from the intention in both directions. In the volitional condition, however, his performance is strongly correlated with intention in both directions, compounding to a highly significant split in the terminal scores. Note that the scales of achievement and the general patterns of the two signatures are virtually antisymmetric, resulting in an apparently null effect for the combined total data base (Fig. II-9[c]).

Figure II-10 shows a similar stark difference in performance by another operator when using the manual mode, wherein

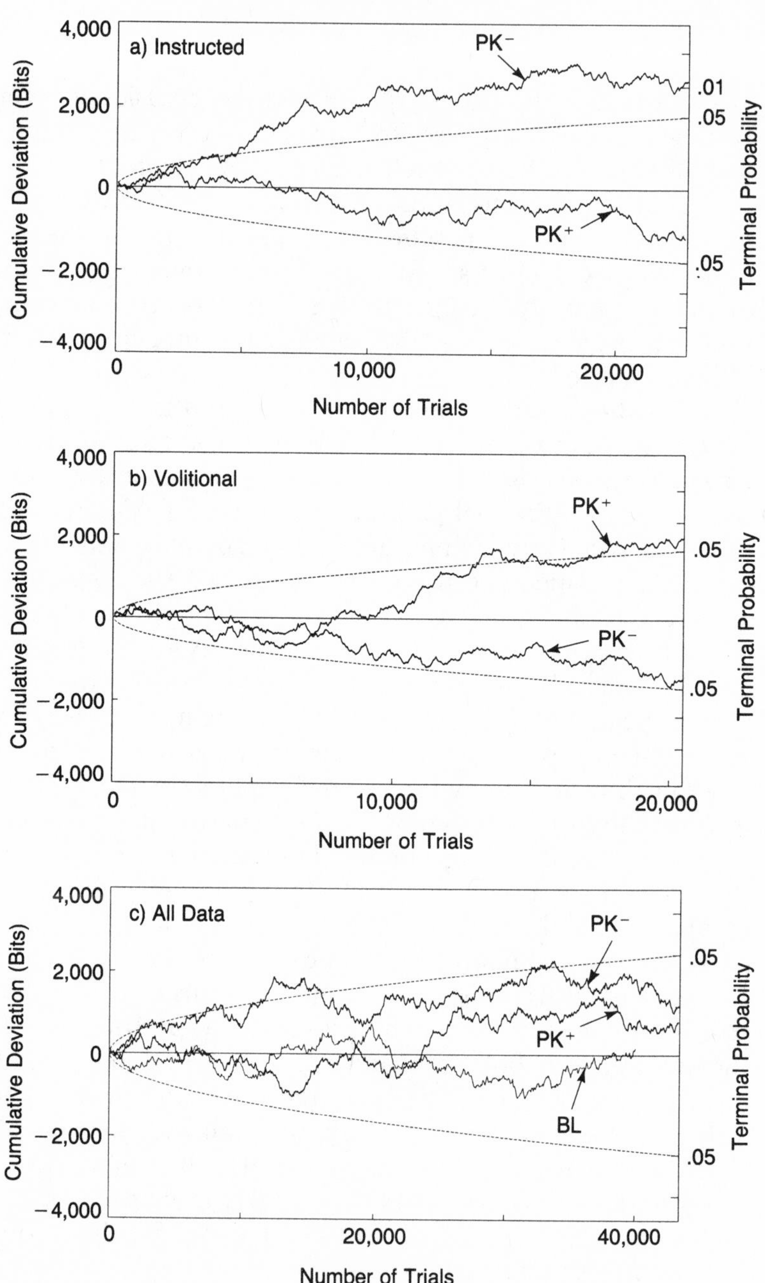

Figs. II-9(a–c) *REG Signature Sensitivity to Volitional/Instructed Option: Operator 55*

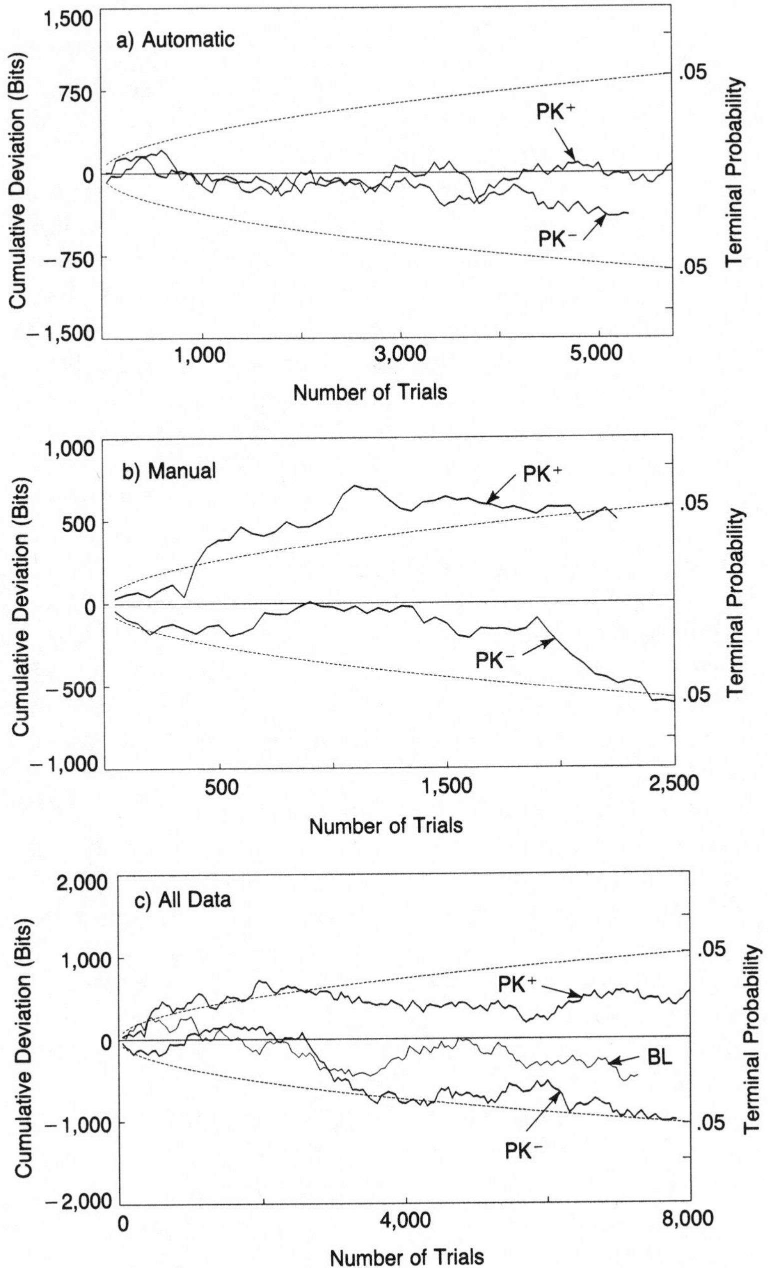

Figs. II-10(a–c) *REG Signature Sensitivity to Automatic/Manual Option: Operator 14*

the initiation switch must be pushed to start each trial, versus the automatic mode, wherein a single push of the switch initiates a regularly spaced sequence of fifty trials.

To interpret these two examples, one might hypothesize that some aspect of the operator's psyche is resistant to the prevailing constraints, namely the imposition of the direction of effort or the unrelenting presentation of new trials in the automatic sequence. However, for other operators just the reverse patterns are observed: strong correlation on instructed or automatic modes, lack of correlation on volitional or manual, from which one might be tempted to conclude that those particular operators prefer not to choose their directions of effort, or the time to initiate each trial. Beyond these disparities, there are other examples where such options seem to have very little effect at all on the operator's performance signature. Such is the case for the operator whose results were illustrated in Figs. II-4 and II-5. In sum, it appears that the influence of these parameters on operator performance is neither qualitatively or quantitatively uniform. For some operators these are unimportant; for others they are consequential, but in different ways and to various degrees.

A similar pattern of individual dependence seems to prevail for such technical parameters as the number of bits per trial, the sampling rate, or the number of trials per run. Although the multivaried options for each of these preclude any complete survey of all possible permutations, selective explorations by several of our operators indicate once again that any given parameter may or may not consistently influence any individual signature, but no generic pattern across these operators has been found.

On the one hand, this lack of general correlation of technical parameters with the observed anomalies complicates considerably the design of experiments and the search for comprehensive explanation. On the other hand, the demonstrated specificity of influence of these parameters on individual operator performance enhances the credibility of the phenomena and raises intriguing possibilities of their ultimate correlation with some psychological or physiological characteristics of the operator. In a sense, the individual operator signatures and their dependence on such parameters may constitute some sort of complex and subtle psychic "Rorschach" that can reveal features of the operator's consciousness not addressed in other venues.

4

BASELINE BIND

Superficially, one would expect that the data generated by the operators in the null intention baseline condition would simply replicate the calibration data that conform so well to the theoretically expected count distributions. And indeed, without exception, the means of all of the baseline series data in this experiment remain within the statistical chance expectations. Yet even here a somewhat more subtle, but at least as intriguing anomaly can be observed. Namely, of the 76 baseline series performed, 7 or 8 of the means would be expected to exceed the .05 terminal probability criterion, in one direction or the other, simply by chance. In fact, not one of them does. Indeed, if all of these baseline terminal *z*-scores are plotted as a histogram and are then fitted by a Gaussian envelope, they form a distribution with its mean almost precisely on the chance value, but with its variance significantly *narrower* than the theoretical expectation (cf. Fig. II-11[a]). Curious as this result may appear, similar effects have been reported elsewhere in a variety of contexts.[4,5]

In contrast, when a similar exercise is performed for all of the PK^+ and PK^- series scores, the best fit distributions have their means shifted appropriately to the terminal values seen in the total operator traces of Fig. II-7(a), but the variances are significantly *wider* than the theoretical values (Figs. II-11[b,c]). Recalling that all of these data were generated under identical conditions, save for the stated directional intention

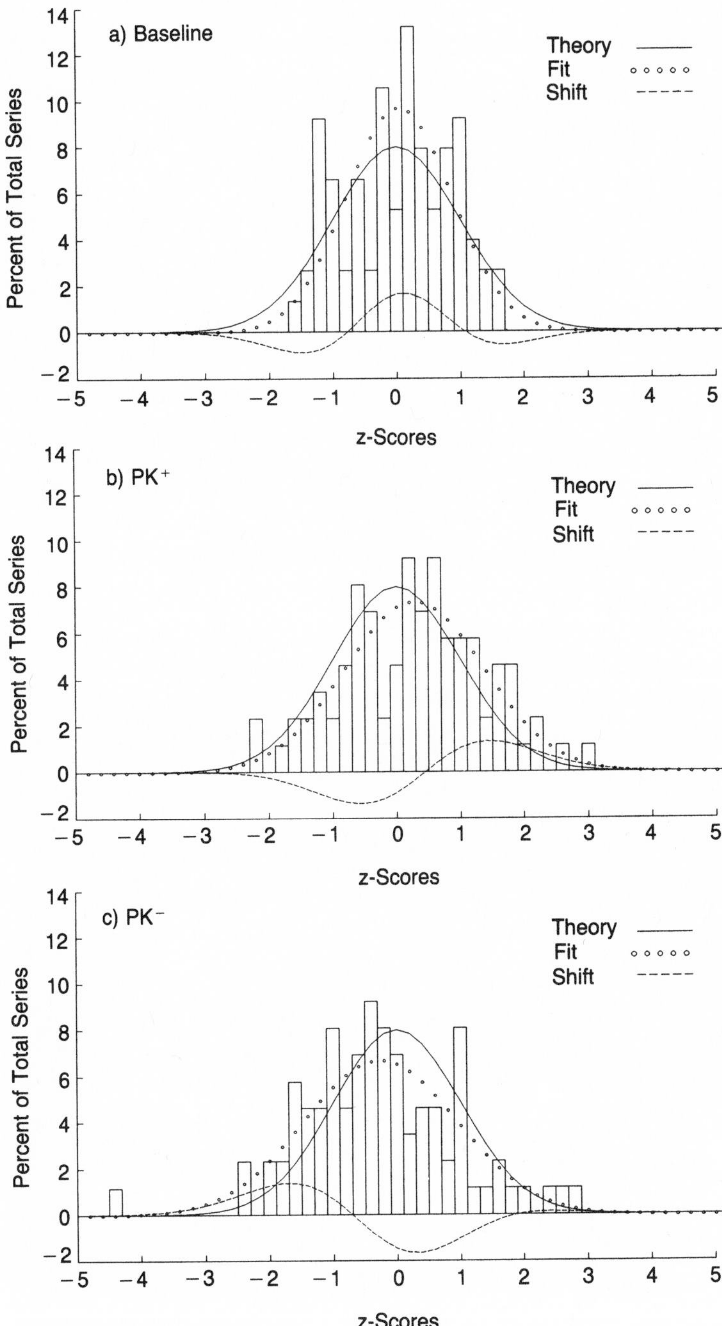

Figs. II-11(a–c) *REG Series z-Score Histograms: BL, PK⁺, PK⁻*

of the operator, it appears that whatever mechanism is effecting the PK^+ and PK^- mean shifts is being inverted to constrict the distribution of baseline scores excessively close to the theoretical mean value. In other words, some conscious or unconscious motivation on the part of the operators to achieve a "good" baseline seems to manifest itself in restraint of series scores from particularly high or low values. So pervasive is this effect that it can be found even in the baseline data of operators showing otherwise chance or even inverted PK results. Interestingly, the effect is not detected in the individual trial scores themselves, or in the calibration data taken independently of the operator experimental series, suggesting that the ultimate accountability of the tripolar series performance may be a pertinent factor.

Such series score histograms can display another fascinating and suggestive feature of the data. Namely, if all of the PK^+, PK^-, and BL series scores are recombined into a balanced mixture, the resulting total histogram returns to a well-behaved Gaussian form, centered on the chance mean and having the predicted variance (Fig. II-12). One intriguing interpretation of this behavior is that the entire family of anomalous effects,

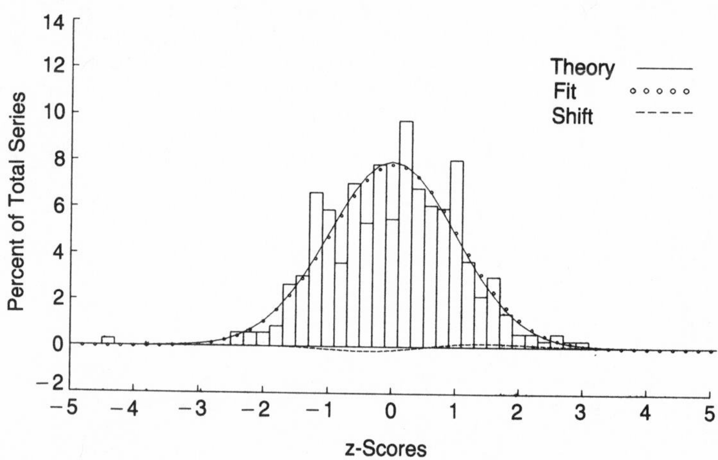

Fig. II-12 *REG Series z-Score Histograms: Balanced Recombination of BL, PK^+, PK^-*

including PK^+, PK^-, and BL, as compounded over the entire base of operators and conditions, can be regarded as simply a selective separation of some grand Gaussian distribution of possible series scores into subsets that are preferentially weighted to higher values (PK^+), lower values (PK^-), and values constricted near the chance mean (BL). In other words, without deforming the global chance behavior of their total results, the operators somehow dispose the system to correlate subsets of the output scores with their intentions. This process bears some similarity to the "Maxwell Demon" paradox of the kinetic theory of gases that is precluded by the second law of thermodynamics. In classical physical theory, no completely isolated system can spontaneously reduce its own randomness or "entropy." Thus any such preferential sorting of scores would itself constitute an anomaly. The possible role of consciousness as an entropy-reducing agent has been a topic of speculation in contexts other than this, and will be addressed further in Section V.

5

PSEUDO SERIES

Experiments such as these inevitably raise the question of the focus of the interaction between the consciousness of the operator and the machine. Is the physical behavior of the electronic noise source being affected, and if so in what way? Or, is the effect possibly more systemic, manifesting itself in the output data in an anomalous statistical form, without alteration of any specific physical process? The obvious first strategy for addressing this question is to replace the noise source unit by others and compare the results. Several similar microelectronic noise units have indeed been substituted for the original source, with no discernible consequences on the overall pattern of results.

To explore this issue more aggressively, a categorically different noise source has been developed that employs microelectronic circuitry to construct signals which, although totally determinate in character, can provide arbitrarily good approximations to theoretical randomness. Such sources are generically termed "pseudorandom." Our particular form uses an array of thirty-one microelectronic shift registers interacting to produce a sequence of two billion pseudorandom bits that does not repeat itself for about sixty hours of continuous operation. This source can be switched into the standard REG apparatus to replace the noise diode and its conditioning circuitry, leaving all attendant processing and display equipment identical. From the perspective of the operator, the two systems are indistinguishable in operation, ambience, and protocol, yet one is ultimately predetermined, the other truly random.

The results obtained using this device are strikingly similar to those achieved with the random source. At this writing, ten operators have completed twenty-nine experimental series that concatenate to the cumulative deviation graph shown in Fig. II-13, having terminal probabilities against chance for PK$^+$ of .081, PK$^-$ of .005, and ΔPK of .003. Five of the series were statistically significant in PK$^+$, five in PK$^-$, and six in ΔPK, compared to the one or two expected by chance, and five of the ten operators achieved significance over their total data bases. Appendix Table D summarizes the individual operator accomplishments. Note that once again, using criteria of series scores, series score direction correlations, or operator total achievement, the pattern of results substantially exceeds chance expectation, indicating that the basic phenomenon is not necessarily tied to explicitly random physical processes.

Perhaps equally indicative is the clear similarity of several of the individual operator signatures to those obtained on the standard REG, using the same technical parameters. In fact, the individual sensitivities of operators to those parameters

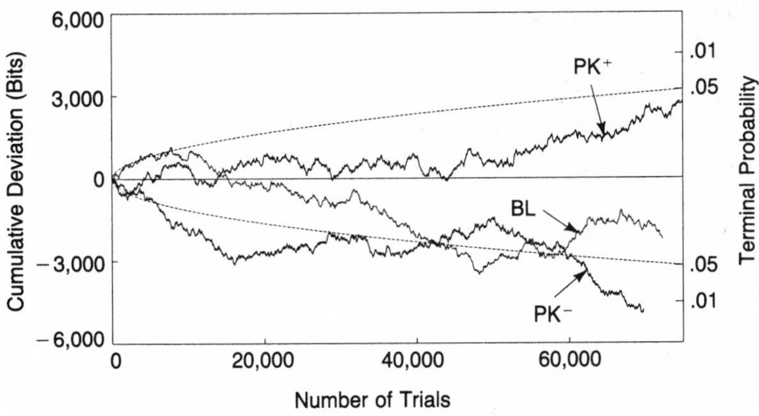

Fig. II-13 *Pseudo-REG Cumulative Deviations: All Operators*

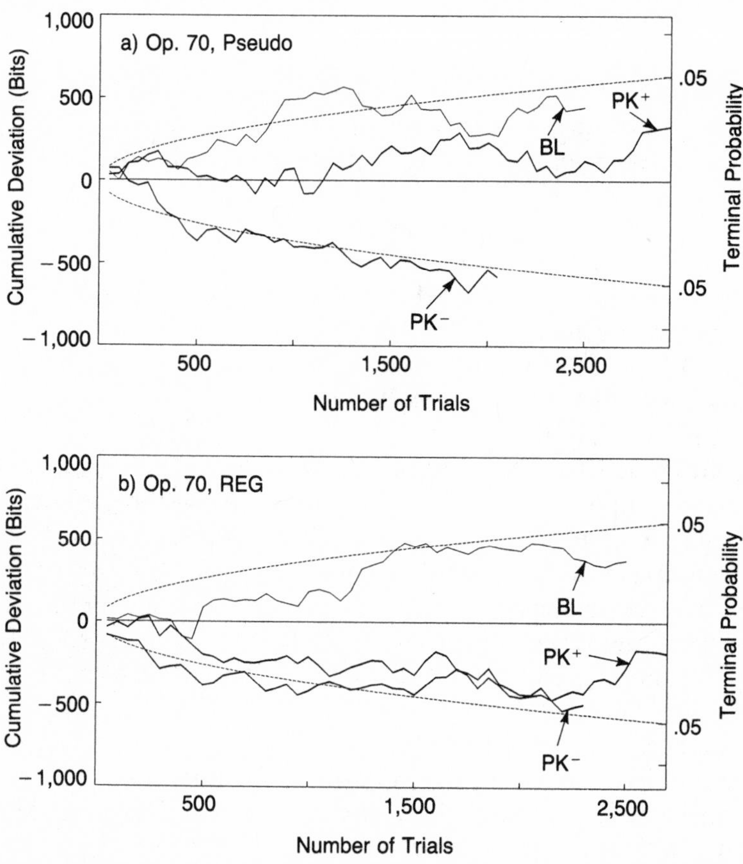

Figs. II-14(a–d) *Comparisons of Pseudo-REG and REG Signatures: Operators 70 and 14*

frequently transfer to Pseudo-REG operation, as well. Figure II-14 shows two comparisons of operator signatures obtained on the two devices. These similarities further incline toward the second of the two hypotheses posed above regarding the mechanics of the anomalous interaction; namely, that the effect seems to function at the systemic informational level, rather than invading the physical processes of any given component. This would also be consistent with the series score segregation model suggested by the histogram of Fig. II-12, which preserved the random integrity of the total data base.

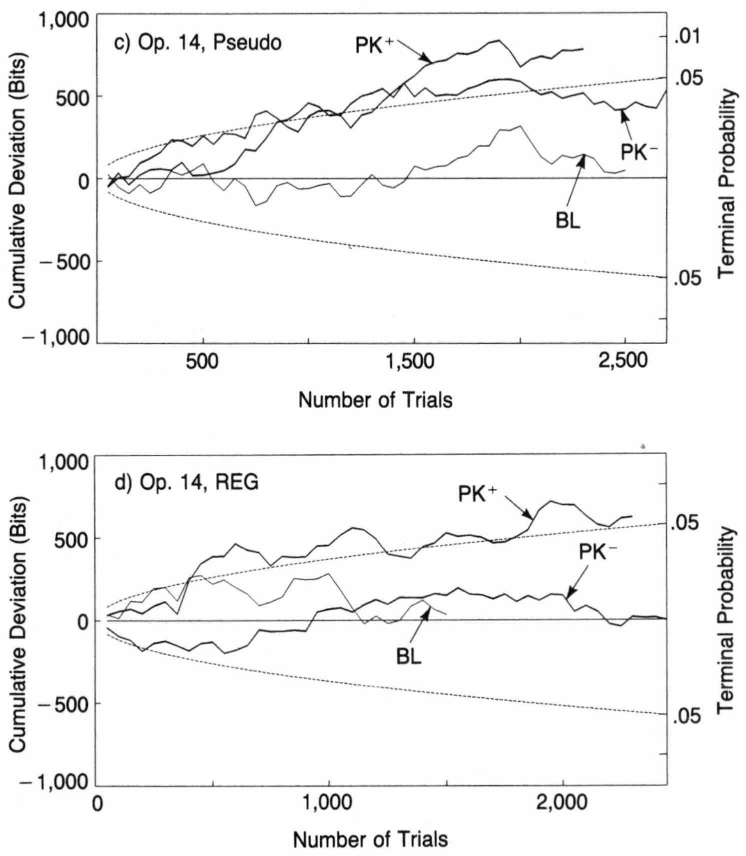
c) Op. 14, Pseudo
PK⁺
PK⁻
BL
1,000
500
0
− 500
− 1,000
500
1,500
2,500
.01
.05
.05
Cumulative Deviation (Bits)
Terminal Probability
Number of Trials
d) Op. 14, REG
PK⁺
PK⁻
BL
1,000
500
0
− 500
− 1,000
0
1,000
2,000
.05
.05
Cumulative Deviation (Bits)
Terminal Probability
Number of Trials

6

MACROSCOPIC MANIFESTATIONS

The strong similarity of the anomalous effects obtained on the randomly driven REG and on the digitally programmed Pseudo-REG prompts broader exploration of this insensitivity of operator signatures to the type of random physical process entailed. Despite the categorical difference in their sources of randomicity, the REG and Pseudo-REG are quite similar in their microelectronic implementation, binary logic, and the extremely small energies involved in their information-processing elements. It is thus reasonable to ask whether such anomalies can also be demonstrated with other than electronic systems, with other than binary statistical systems, or with systems of macroscopic scale.

In 1894, the British eugenicist Francis Galton described a mechanical apparatus that could illustrate the development of random Gaussian distributions by the compounding of a multitude of binary events:

> It is a frame glazed in front, leaving a depth of about a quarter of an inch behind the glass. Strips are placed in the upper part to act as a funnel. Below the outlet of the funnel stand a succession of rows of pins stuck squarely into the backboard, and below these again are a series of vertical compartments. A charge of small shot is inclosed. When the frame is held topsy-turvy, all the shot runs to the upper end; then, when it is turned back into its working position, the desired action commences. . . . The shot passes through the funnel and issuing from its narrow end, scampers

> deviously down through the pins in a curious and interesting way; each of them darting a step to the right or left, as the case may be, every time it strikes a pin. The pins are disposed in a quincunx fashion, so that every descending shot strikes against a pin in each successive row. The cascade issuing from the funnel broadens as it descends, and, at length, every shot finds itself caught in a compartment immediately after freeing itself from the last row of pins. The outline of the columns of shot that accumulate in the successive compartments approximates to the Curve of Frequency, and is closely of the same shape however often the experiment is repeated. . . .
>
> 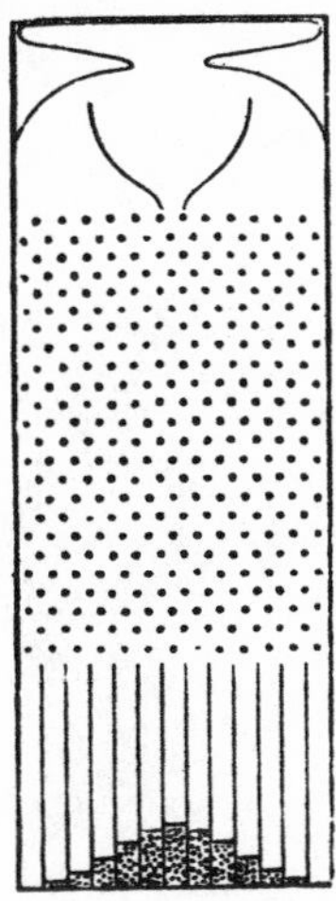
>
> The principle on which the action of the apparatus depends is, that a number of small and independent accidents befall each shot in its career. In rare cases, a long run of luck continues to favour the course of a particular shot towards either outside place, but in the large majority of instances the number of accidents that cause Deviation to the right, balance in a greater or less degree those that cause Deviation to the left. Therefore most of the shot finds its way into the compartments that are situated near to a perpendicular line drawn from the outlet of the funnel, and the Frequency with which shots stray to different distances to the right or left of that line diminishes in a much faster ratio than those distances increase. This illustrates and explains the reason why mediocrity is so common.[6]

This kind of device has come to be known as "Galton's desk," and various versions of it may be found in science museums and instructional laboratories throughout the world.

A form of this machine tailored to our particular purposes has also been constructed in our laboratory. Called the "Random Mechanical Cascade" (RMC), this apparatus is ten feet high and six feet wide, and is mounted on a wall facing a comfortable couch. In operation, 9000 polystyrene balls, ¾″ in diameter, trickle downward from an entrance funnel into a

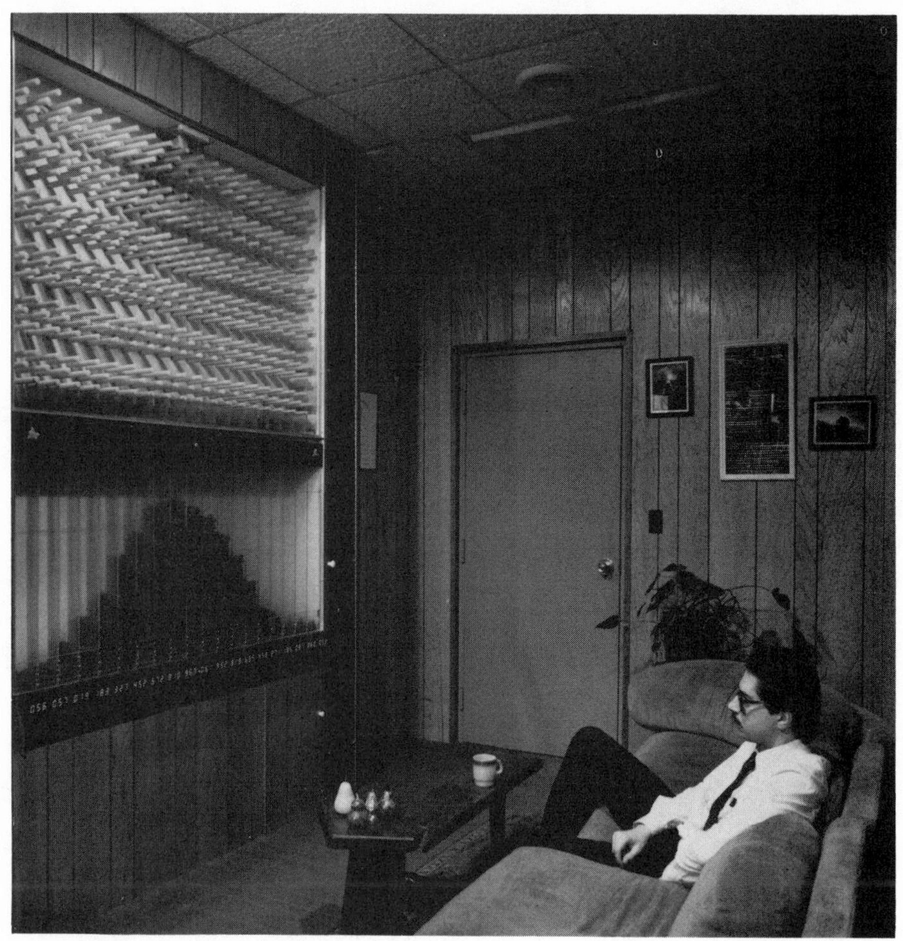

Random Mechanical Cascade Apparatus

quincunx array of 330 nylon pegs, also of ¾″ diameter and mounted on 2¼″ centers. The balls bounce in complex random paths through the array, colliding elastically with the pegs and other balls, accumulating finally in nineteen parallel collecting bins across the bottom. The front of the peg chamber and of the collecting bins below it are transparent so that the cascade of balls and their developing distribution of bin populations are visible as feedback to the operator. By appropriate combinations of peg spacing, ball inlet arrangement, and material properties, the resulting distribution of ball populations in the collecting bins can be brought to a good approximation of a Gaussian distribution (cf. Fig. II-15).

The entrance to each collecting bin is equipped with a photoelectric sensor that detects and records the arrival of each ball, and the growing populations of all bins are displayed via LED counters below each bin and graphically on a computer terminal screen. The disposition of each of the 9000 balls in every run is recorded on-line in an appropriately coded computer file that can later be accessed to yield a faithful reproduction of the complete history of all of the bin fillings for more detailed study or to calculate statistical properties of the terminal distribution. In addition, a Polaroid photograph of the distribution and bin counts is taken after every run. As with the REG devices, extensive calibrations are accumulated to provide background statistical data and to explore possible sensitivities to temperature and humidity, which are routinely measured and recorded before each run.

The experimental RMC protocol calls for the operator, seated on the couch approximately eight feet from the machine, to attempt to distort the distribution of balls in the bins toward

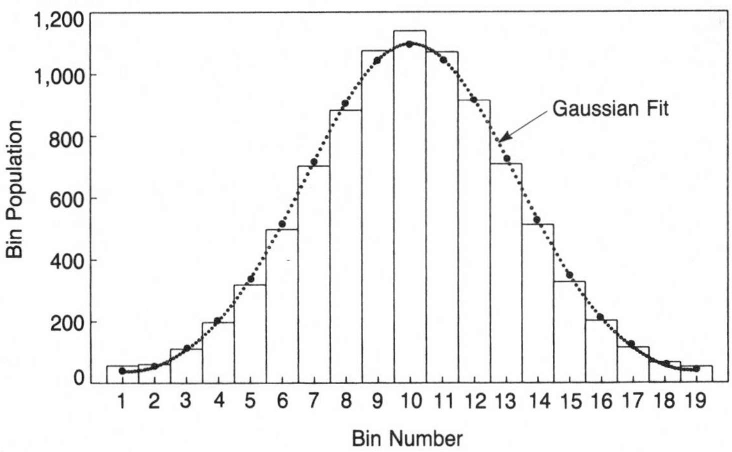

Fig. II-15 *RMC Average Bin Populations: 952 Baseline Runs*

the right or higher numbered bins (PK^+), or to the left or lower numbered bins (PK^-), or to generate baselines (BL). These efforts are interspersed in concomitant sets of three runs, lasting approximately twelve minutes each, again providing tripolar strings of data. Unlike the REG, where a theoretical baseline, confirmed by calibration, is available for comparison with the operator-generated data, the internal mechanics of the RMC are too complex to submit to detailed theoretical prediction. Rather, statistical treatment of the data is based upon comparison of the means of the PK^+ and PK^- distributions with the local baseline of the same experimental set. This strategy has the additional benefit of minimizing the effects of any long-term drift in machine characteristics that could arise from ball or pin wear or other mechanical or environmental influences.

The results of these RMC experiments, like those of the REG, also display clear anomalies and strongly operator-specific patterns of achievement. Figures II-16(a–d) show a few typical cumulative deviation signatures. All of the individual operator performances are tabulated in Appendix Table B. At this writing, 22 operators have completed a total of 76 experimental series, each consisting of 10 (or 20) runs under each of the three intentions. Of these, 8 operators, or 36%, have generated

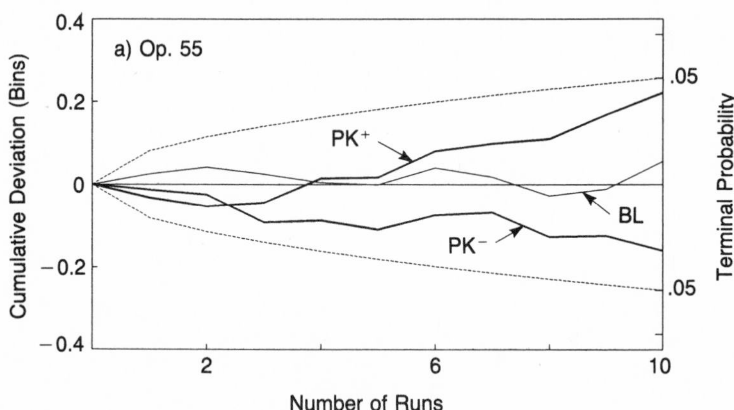

Figs. II-16(a–d) *RMC Cumulative Deviation Signatures: Four Operators*

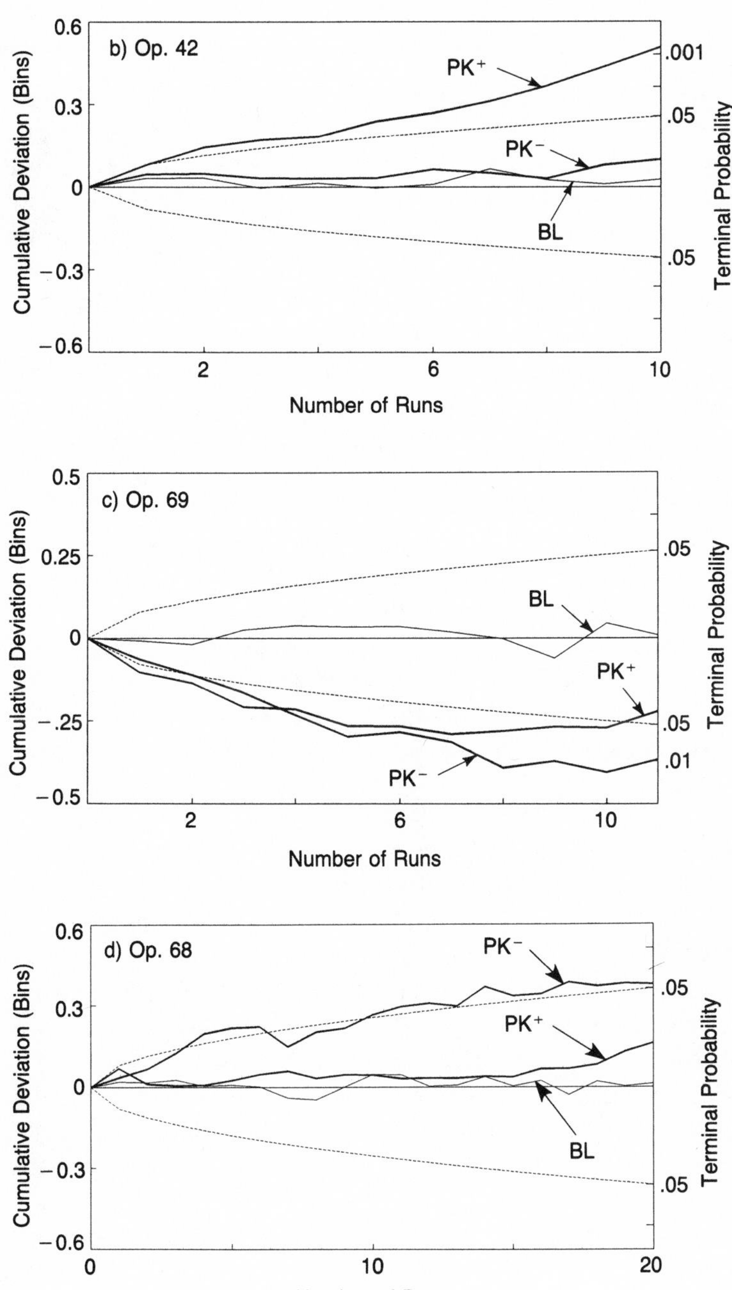

b) Op. 42
PK⁺
PK⁻
BL
Cumulative Deviation (Bins)
Terminal Probability
.001
.05
.05
Number of Runs
c) Op. 69
BL
PK⁺
PK⁻
Cumulative Deviation (Bins)
Terminal Probability
.05
.05
.01
Number of Runs
d) Op. 68
PK⁻
PK⁺
BL
Cumulative Deviation (Bins)
Terminal Probability
.05
.05
Number of Runs

significant overall data bases: two with significant PK^+ effects; three with significant PK^-effects; and three with significant separation of PK^+ and PK^-. Four of the 76 series are significant in the PK^+ direction, 14 in the PK^- direction, and 13 show significant ΔPK splits, compared to the chance expectation of three or four.

The overall data base is also statistically significant, but displays a curious asymmetry in the PK^- direction. This is primarily attributable to a large contribution by one operator with a very strong PK^- effect and null PK^+. Figure II-17(a) shows the entire data base, Fig. II-17(b) displays the data base with this operator omitted, and Fig. II-17(c) concatenates the first ten runs of all operators.

As in the REG and Pseudo-REG experiments, the total number of series conforming to the intended direction to any degree is considerably higher than chance prediction. Compared to the 38 of each expected by chance, 44 of the PK^+ efforts (58%), 47 of the PK^- efforts (62%), and 53 of the ΔPK separations (70%) agree with the operators' stated intentions. Thus, once again we find a pattern of small but numerous shifts of the mean compounding to an overall accumulation of significant anomaly. We also again find operator-specific dependencies of the results on the secondary parameters of the experiment, much like those described for REG.[7]

Perhaps of highest importance, however, is the continuing similarity of several individual operator signatures with those demonstrated in the microscopic REG experiments. Figure II-18 compares signatures on the REG and RMC devices for two operators who have completed three series on each. Figure II-19 compares signatures of another operator with substantial data bases on all three REG, Pseudo-REG, and RMC experiments. Despite their inherently stochastic character, the evident gross similarities in the configuration of these traces has major implications for experimental design and theoretical modeling: namely, although the observed anomalous effects are clearly operator-specific and in many cases condition-specific, they appear not to be nearly so device-specific, over a wide range of physical processes, scales, and energies. Thus, once again,

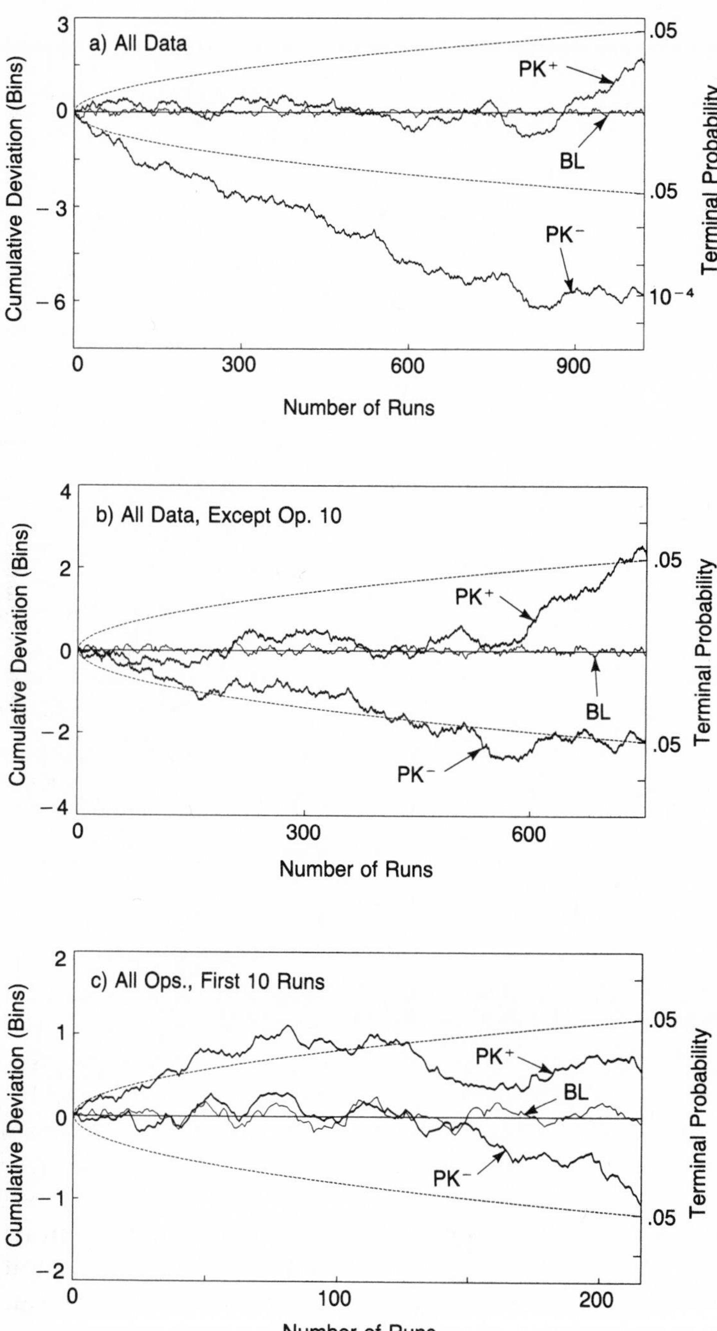

Figs. II-17(a–c) *RMC Grand Cumulative Deviations: All Operators*

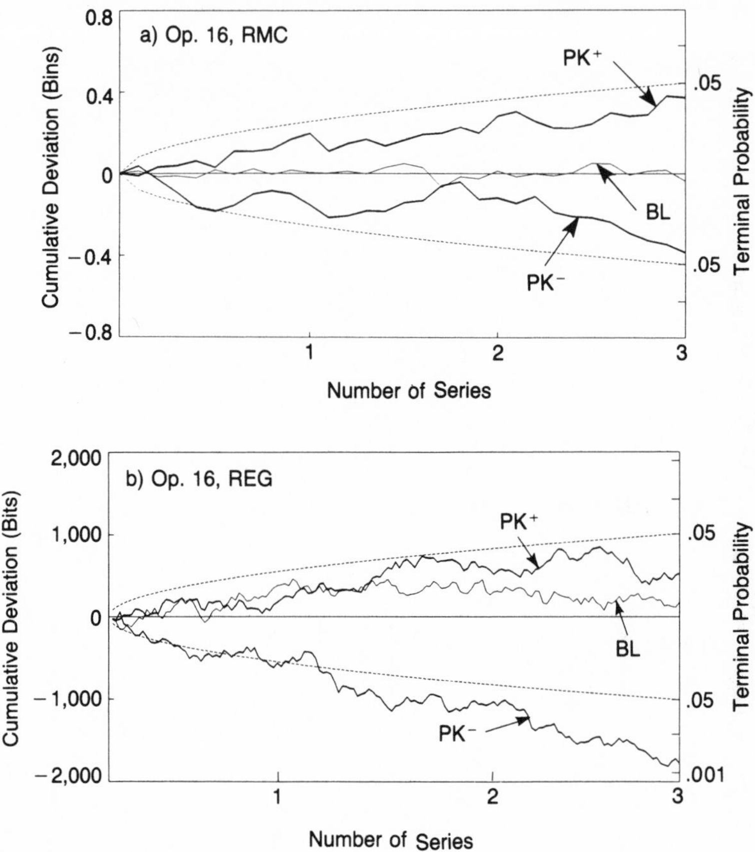

Figs. II-18(a–d) *Comparisons of RMC and REG Signatures: Operators 16 and 94*

any direct influence of operator consciousness on particular physical processes, such as the flow of electrons in the REG noise diode, the point of incursion into the deterministic Pseudo-REG bit string, or the cascade of balls in the RMC, appears to be less likely than some more holistic interaction with the statistical information common to all three of these systems.

Although the RMC differs substantially from the REG in its scale and physical process, it retains a certain quasi-digital character in the manner in which it generates information. Specifically, each falling ball, upon collision with a peg, may be diverted either to the right or to the left, and it is the

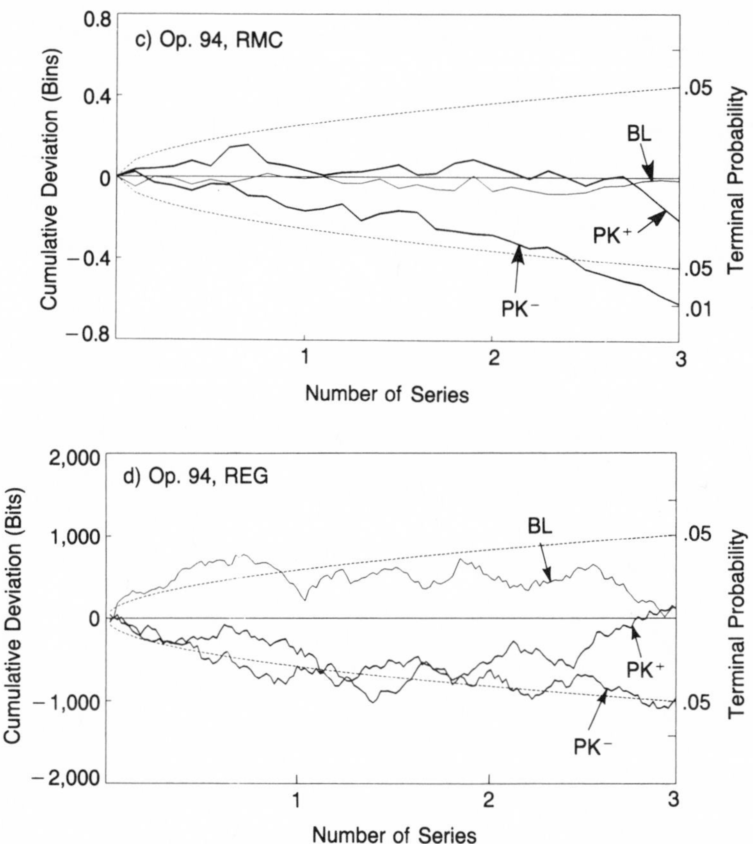

compounding of these binary right/left options that primarily determines the terminal distribution in the bins. To be sure, in this machine the binary right/left probabilities are not simply .50/.50, nor are the balls' subsequent trajectories uniform, but nonetheless, a synthetic binary quantification can be imposed on the analysis. A further step in exploring operator-related anomalies, therefore, would be to move to some physical system that is more analogue in character, in the sense that its central random process and output lend themselves to continuum representation.

A number of such analogue experiments are being pursued in our laboratory. Some of these are optical in character, others

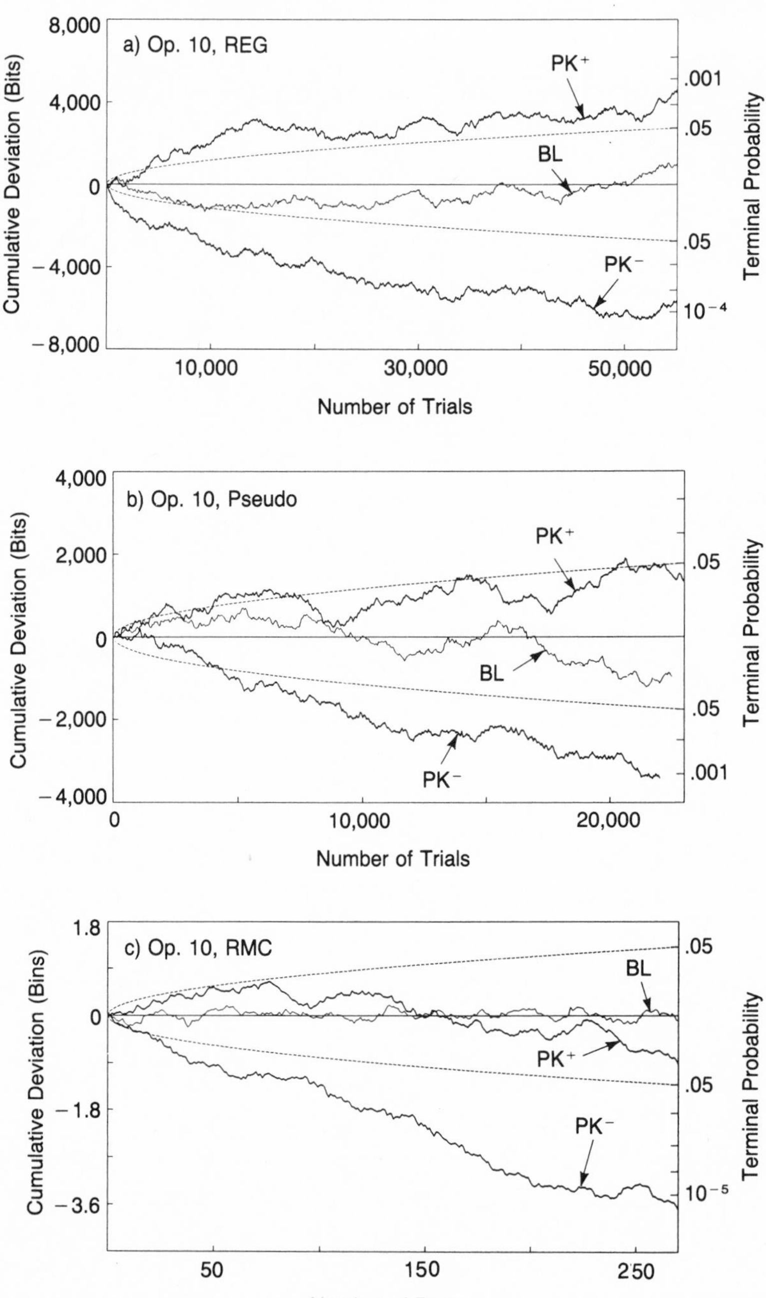

Figs. II-19(a–c) *Comparisons of REG, Pseudo-REG, and RMC Signatures: Operator 10*

thermal, mechanical, fluid dynamical, or electrical. Each presents continuous feedback to the operator indicative of the progress of some continuous physical property of the device or system, such as the distance between two interferometer mirrors, the temperature of a small gas cavity, the strain in a solid plastic bar, the period of a linear pendulum, the amplitude of hydrodynamic wave crests, or the location of striations in a glow discharge. All of these experiments are approached via the same tripolar protocol followed for their digital counterparts, but the reduction of their analogue data in quantitative statistical terms is more complex. Although suggestive effects have also been observed in these experiments, our results so far are too provisional to justify discussion.

7

SUPERPOSITION AND SEPARATION

Our sixth question regarding operator-related anomalies concerned the combination of more than one operator's efforts on a given device and task. For example, it is reasonable to ask whether two operators showing comparable signatures may, in a concerted effort, reinforce one another to produce yet more substantial effects, or whether operators of dissimilar propensities may tend to cancel one another's influence. A modest program of experiments addressing this superposition issue is currently in progress on the REG and RMC machines. While the accumulated data are much less extensive than for the single operators, some general characteristics can already be identified. Namely, it appears that anomalous results can occur with similar frequency to those of the individual efforts, but in patterns that are not simply linear combinations of the separate signatures. Figures II-20(a–f) compare two sets of individual signatures with the corresponding composite efforts. Although the conduct and interpretation of these co-operator experiments are more complex, such superposition properties are important ingredients in the development and application of the theoretical model we shall propose in Section IV.

The dependence of the signatures on the distance of the operator from the apparatus is also of prime interest, both experimentally and theoretically. In the next section we shall describe an experimental and analytical study of the anomalous

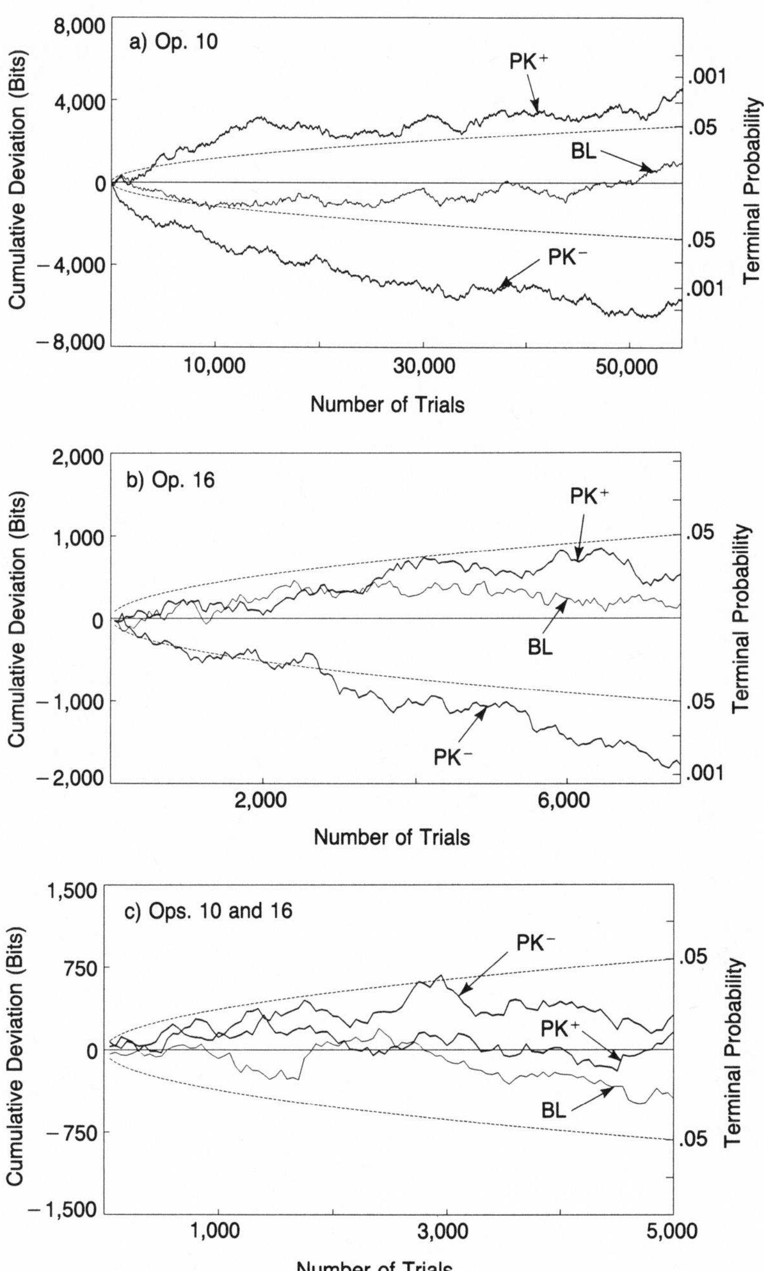

Figs. II-20(a–c) *Comparisons of Individual and Composite REG Signatures: Operators 10 and 16*

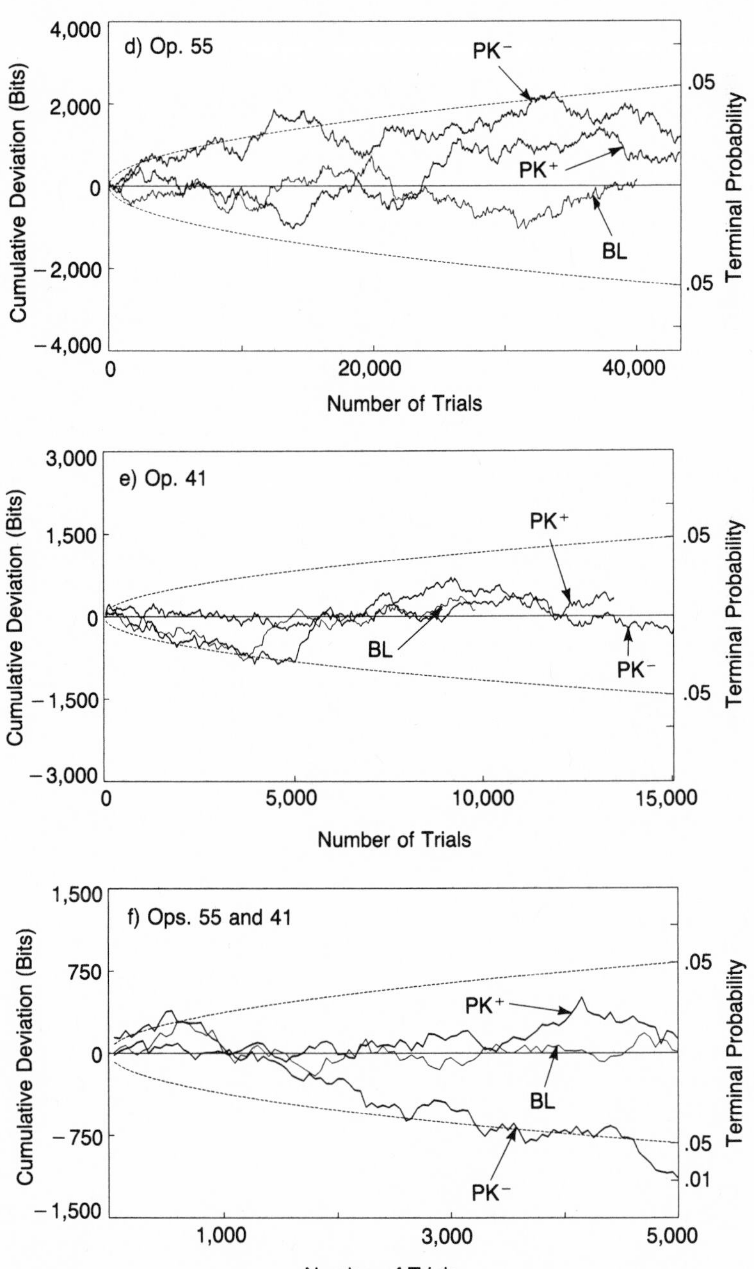

Figs. II-20(d–f) *Comparisons of Individual and Composite REG Signatures: Operators 55 and 41*

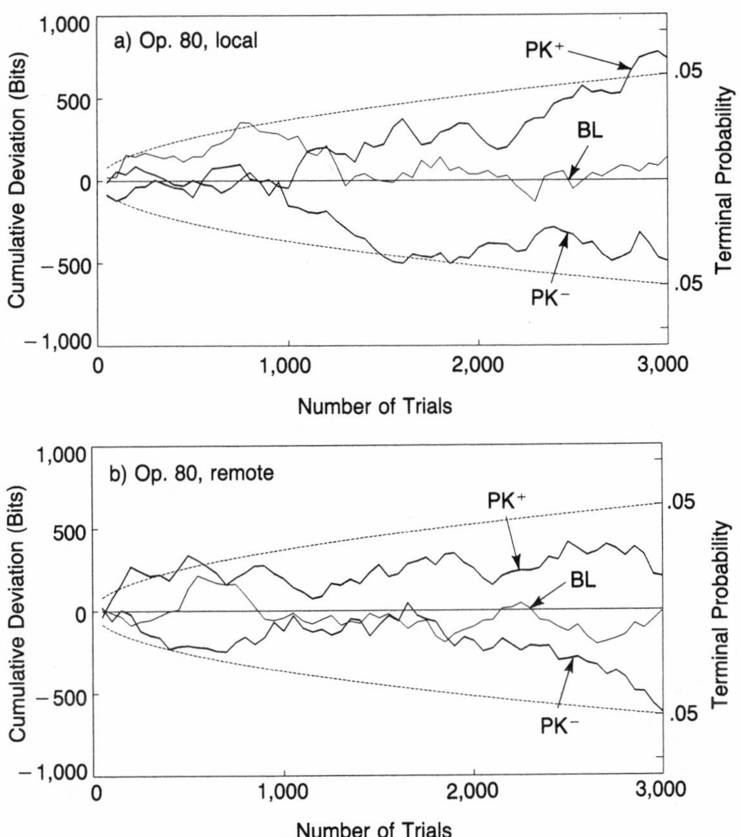

Figs. II-21 (a, b) *Comparisons of Local and Remote REG Signatures: Operator 80*

acquisition of information about geographic targets remote in space and time. The success of these experiments raises the possibility that man/machine anomalies may also be achieved without the operator physically proximate to the device. To study this, a separate program of experiments has been established in which given operators address the REG and RMC tasks from various remote locations, ranging from a few miles to global distances. As with the co-operator experiments, various anomalous signatures continue to be observed, some of which are consistent with the local efforts, and some of which show characteristic differences. Figures II-21(a–d) show com-

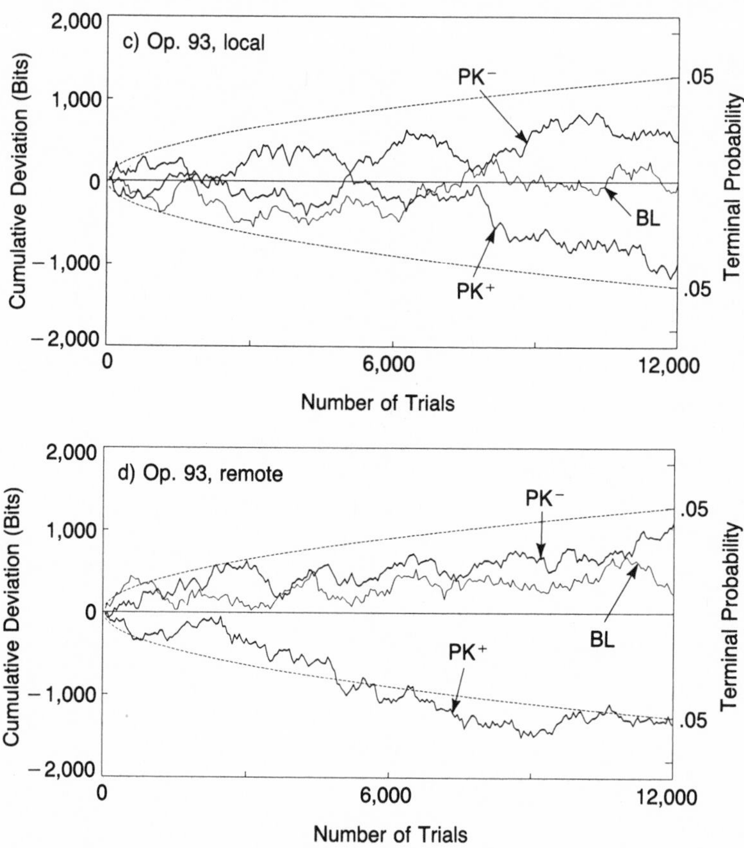

Figs. II-21(c, d) *Comparisons of Local and Remote REG Signatures: Operator 93*

parisons of remote and proximate REG signatures for two of our more prolific operators, stationed in upper New York State and Texas, respectively. We cannot claim at this point to understand the factors bearing on such performance, let alone to assess the distance variable in any comprehensive way. Nonetheless, even qualitative observation of such long-distance effects raises severe implications for the ultimate representation, comprehension, and application of the basic process involved.

3

PSYCHOLOGICAL PARAMETERS

The final item on our original list of queries addressed the correlation of operator performance with psychological and physiological characteristics, and on this point, at present, we have least to say. For a number of reasons, including operator comfort and spontaneity, the overall capacity of the program, the limited number of operators contributing to the data base, the questionable efficacy of conventional indicators and tests, and the substantial activity on this front at other laboratories, we have so far attempted no formal codification of operator personality traits, styles of operation, or physiological characteristics for correlation with their signatures of achievement. This does not imply any discounting of these influences; to the contrary, both our empirical data and our theoretical model make it clear that at least some of these factors are relevant.

On the basis of informal discussions with our operators, casual observations of their styles, occasional remarks they record in the experimental logbooks, and our own experiences as operators, it is clear that individual strategies vary widely. Some operators self-impose preliminary meditation exercises, employ visualization techniques, or attempt to identify with the device or process in some transpersonal context. Others invoke competitive strategies, attempting to outperform other operators, their own earlier results, or simply the laws of chance. Some operators consistently utilize the same feedback mode throughout, others alternate among the various feedback options in some regular or irregular pattern, and a few ignore them entirely. Many of the operators attribute anthropomorphic characteristics to the devices, and frequently engage in various forms of exhortation, coaxing, pleading, or threat. Unfortunately, there is little pattern of correlation of such strategies with achievement. Rather, it appears that such tactics are also operator-specific and often transitory; what works well for one does not necessarily help another, and what works on one occasion may fail on the next.

If there is any unity in this diversity of strategy, it would be that most effective operators seem to associate successful performance with the attainment of some sense of "resonance" with the device. This has been variously described in such terms as:

> ". . . a state of immersion in the process which leads to a loss of awareness of myself and the immediate surroundings, similar to the experience of being absorbed in a game, book, theatrical performance, or some creative occupation."
>
> "I don't feel any direct control over the device, more like a marginal influence when I'm in resonance with the machine. It's like being in a canoe; when it goes where I want, I flow with it. When it doesn't, I try to break the flow and give it a chance to get back in resonance with me."

The concept of resonance will also assume major importance in the formulation of our theoretical model, and will be readdressed in that context.

Some of the experimental data, such as those displaying extra-chance results in directions opposite to intention, those displaying sensitivity to volitional/instructed or manual/automatic options, or those displaying constriction of baseline series scores, suggest that unconscious or subconscious expectations or aversions play at least as important a role as conscious intention in determining the effects. Certainly much of the parapsychological research literature supports the proposition that far more than superficial psychological processes are involved. Unfortunately, few of the existing psychological tests are likely to be effective in correlating such subtle aspects, let alone in comprehending or incorporating them into a phenomenological model. Our own informal explorations of the more obvious psychological correlates, such as sex, handedness, analytical versus aesthetic personality, assertive versus restrained interactive style, and systematic versus impulsive approaches to the tasks, have indicated little relationship with demonstrated or predicted success.

Physiological monitoring has also been eschewed in this program for similar reasons: none of the readily accessible physiological parameters holds much promise of being fundamentally relevant, and such testing inevitably places the operator in the role of "subject," or object of the study. This, in turn, can affect his sense of purpose, freedom of style, and relationship to the laboratory staff and program goals. Without disputing the possibility that correlations between physiological indicators and operator performance may exist, on balance we have deemed it better to regard our operators as participating members of the research team, rather than as objects of clinical study.

9

OVERALL OBSERVATIONS

The complete body of experimental data acquired in this laboratory over the past eight years on the several REG and RMC experiments outlined above compounds to clear evidence of a legitimate scientific anomaly. While small segments of these results might reasonably be discounted as falling too close to chance behavior or displaying yields too marginal to justify revision of prevailing technical or scientific tenets, taken in concert the entire ensemble establishes an incontrovertible aberration of substantial proportions. The case can equally well rest on the overall data bases of certain individual operators. For example, Fig. II-22 and Appendix Table C summarize the results of nearly 100 experimental series generated by one operator interacting with the REG, Pseudo-REG, and RMC devices over a period of several years, covering a broad range of technical parameters, environmental conditions, and personal moods and strategies. By any reasonable statistical criteria, the likelihood of this concatenation of results occurring by chance is infinitesimally small. Also highly persuasive is the degree of internal consistency of many of the individual operator performances. Whether inferred from qualitative inspection of cumulative deviation graphs or more rigorously established by quantitative statistical assessments such as analysis of variance, regression, or other goodness-of-fit tests, the correlation of specific patterns of results with individual operators is very strong.

The credibility of all of these claims rests on the sizes of their respective data bases. Frequent reference has been made throughout this text and in the illustrations to the large number of trials or bits involved in any particular portion of the experimental program. The point is not so much to proclaim the scale of effort, but to emphasize the overarching importance of basing conclusions on adequately large blocks of results. As any of the cumulative deviation graphs amply illustrate, experiments of this type tend to yield very noisy data wherein a spectrum of random fluctuations can obscure systematic trends. Only after such trends have compounded to displacements that are large compared to the amplitude of the noise oscillations can claims properly be made of their validity and correlations with the prevailing experimental parameters attempted. As just one example, Figs. II-23(a,b) compare cumulative deviation graphs for the first 50-trial tripolar runs of one particular REG series with a similar presentation of 2500-trial data. On the basis of the smaller data set, we might be tempted to conclude that there is no significant PK effect and that, if anything, the baseline is anomalous. However, when the full set is examined, these apparent trends are seen to be transitory features that eventually revert to a substantial and protracted bias toward significant PK results, straddling a well-behaved baseline. Con-

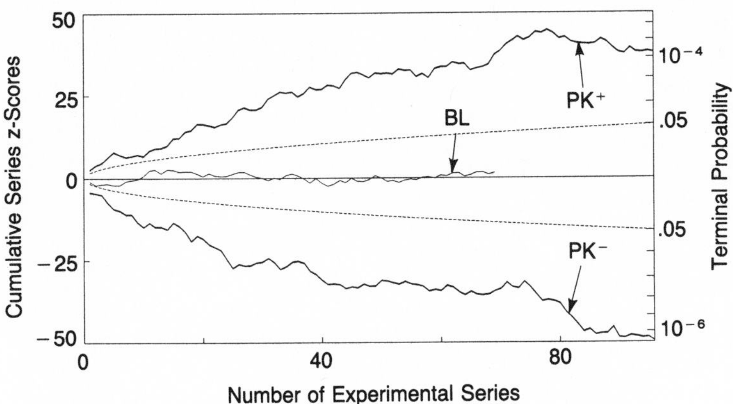

Fig. II-22 *Combined Cumulative Deviation Signature: Operator 10, All Experiments*

versely, examples could be shown where initially promising trends revert to chance behavior as the series progresses. Unfortunately, the literature of this field contains a number of potentially misleading presentations where incorrect inferences have prematurely been drawn from inadequate data bases in just this fashion. That is, the effects claimed are more likely attributable to local variations in the intrinsic noise pattern of the data, rather than to any long-term systematic bias.

The necessity to provide a sufficient block of data to allow systematic trends to emerge clearly beyond the random noise

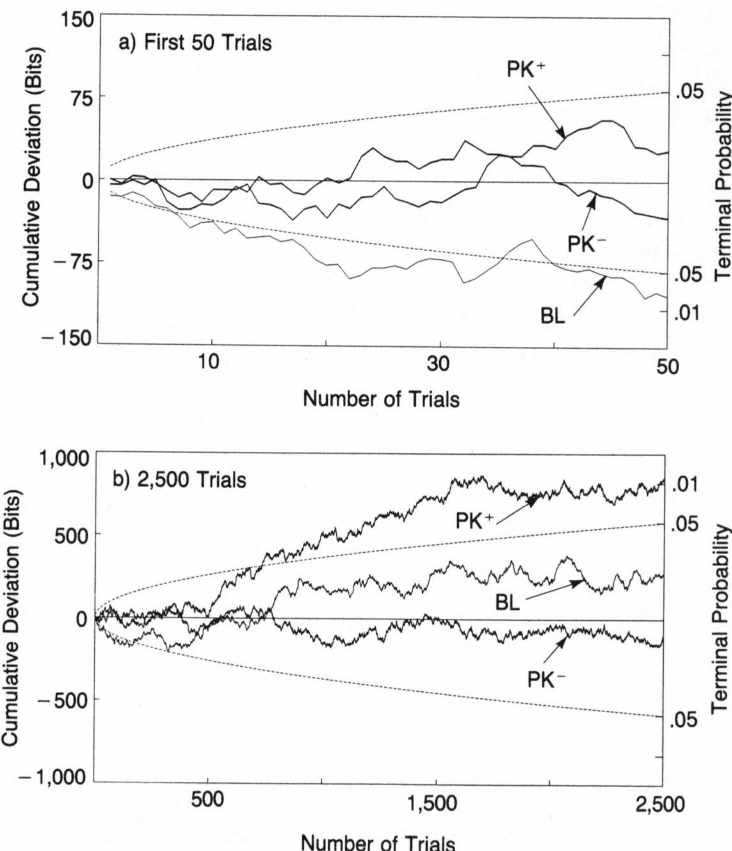

Figs. II-23(a,b) *Cumulative Deviation Graphs: First 50 trials only, compared with full 2500 trials*

dictates the minimum data set sizes, and thereby the inherent data rate capabilities of the experimental equipment and protocols. These, in turn, translate into the technical capacity and endurance of the project and the capital and temporal investments therein. Such technical capacity is now a *sine qua non* of productive experimentation of this class and is the factor underlying our earlier statement that only recently have definitive experiments on these phenomena become technically feasible.

Now, however, given the total size of the individual and collective data bases in hand, their explicit and implicit replicability, the magnitudes of the statistical biases, and their undeniable correlations with intention and with individual operators, it appears that the classical paradigm separating physical process from the conscious and subconscious desires of the human psyche must admit some pragmatic and philosophical problems. On the practical side, since noise diodes, sampling circuits, counters, and display units of the types used in our experiments are common elements in a host of contemporary microelectronic information-processing and control gear, it does not seem unreasonable to question whether such systems, when intimately linked to the processes of human consciousness, are totally immune to aberrations of performance like those displayed here. To be sure, our results have been achieved in very specific contexts, on very sensitive devices, and have failed to implicate any particular technical components; hence, any generic inferences of microelectronic vulnerability may not be warranted at this point. But given the growing engineering reliance on integrated computational circuits and delicately poised microprocessor control systems, some elements of which switch at only a few quanta of energy and tolerate far less binary error than the bits-per-thousand deviations we have observed, a substantial technical problem may conceivably be lurking. For example, the routine reliance on the complete randomicity of microelectronic noise sources for reference and calibration purposes in guidance, control, and communication systems, in view of our results with very similar noise components, now becomes a legitimate concern. In a longer and more positive view, substantial technical opportunities may also be envisaged for enhancing the efficiency and range of application of microelectronic systems by better tuning their characteristics and functions to the consciousness of their human operators. We are far from possessing the basic knowledge to guide conception, research, and development of such man/machine complementarities, but the vision is taking form, and the potentialities are vast.

From a more philosophical perspective, data like those compounded above resemble many major anomalies that have forced redirections of science in the past. In terms of our "two-step" analogy, the experimental foot of science is here encountering an obstacle to its progress that its contemporary theoretical stance cannot surmount, and some reposturing is needed. Certainly, many more experiments will be required fully to define the scale and character of this anomaly. Some of these will be described in the following section; additional data and perspectives will emerge from other laboratories. But the empirical case is already strong enough to warrant reexamination of the prevailing position of science on the role of consciousness in the establishment of physical reality, with the goal of generalizing its theoretical concepts and formalisms to accommodate such consciousness-related effects as normal, rather than anomalous phenomena.

SECTION THREE

Precognitive Remote Perception

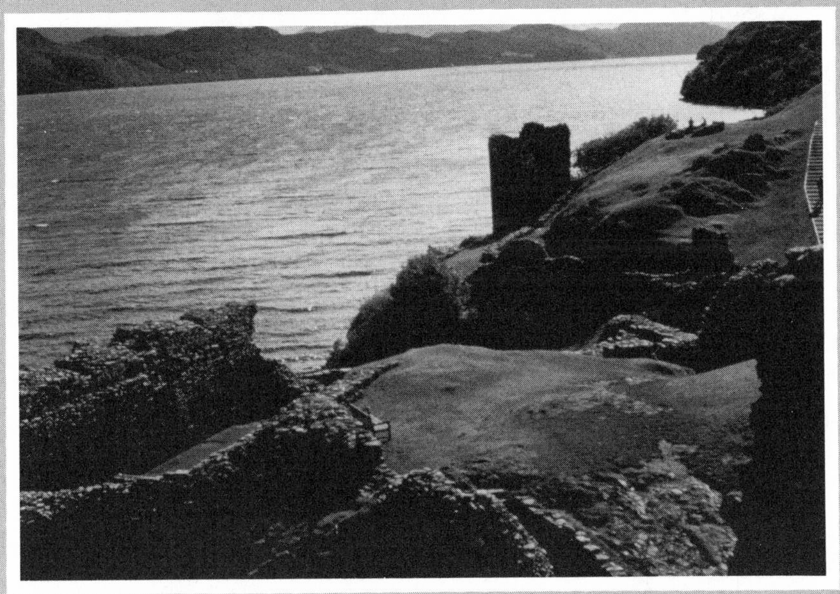

Man also possesses a power by which he may see his friends and the circumstances by which they are surrounded, although such persons may be a thousand miles away from him at that time.

—PARACELSUS[1]

Ruins of Urquardt Castle, Loch Ness, Scotland

1

TV TARGET

On a summer morning in 1977, a television crew from *CBS News Magazine* assembled at Mundelein College in Chicago with a small group of researchers who had for some time been experimenting with a psychic research protocol known as "Remote Viewing." After a brief review of the previously specified logistics, one of the experimenters, whom we shall call "Agent," left in a van with some members of the TV team to drive into the central city. A second, whom we shall call "Percipient," remained behind with the rest of the group and began an attempt to describe where Agent would be an hour and a quarter later. Percipient was comfortably seated in a lounge area throughout the effort and quietly dictated her perception to the camera crew.

After arriving downtown, Agent and her portion of the TV team remained inside their van, parked in the vicinity of the Chicago Tribune building. At the scheduled time, one of the group activated a random number program on a hand calculator to select one of ten envelopes prepared earlier by other members of the CBS staff. Each envelope contained directions to a different target location accessible by a 30-minute drive from their present spot. The target thus randomly selected was the Rockefeller Chapel on the campus of the University of Chicago, to which they then drove.

On arrival at the chapel, Agent and crew disembarked from the van and meandered about the steps and plaza in front of the building and along its sides, videotaping the scene and discussing its features. Their intentions to film the interior were thwarted by a ceremony in progress, but some portions of the vestibule were included. Figures III-1(a–c) show photographs of the exterior and interior of this chapel taken subsequently under more convenient conditions.

A videotape of this experiment, including Agent's and Percipient's behavior and commentary, scenes of the chapel target, and some discussion of the technique and implications of the results by other interested scholars were featured in an episode of *CBS News Magazine* broadcast nationally in January 1978. The full transcript of Percipient's impressions of the target scene, portions of which were also presented in that broadcast, reads as follows:

CBS Interviewer:

[Percipient], it is exactly 1:00 o'clock p.m. We'd like you to visualize where you think [Agent] is going to be at 2:15 this afternoon. She could be anywhere in the entire metropolitan Chicago area. So just relax and very carefully try and visualize where she is going to be. There's a pencil and paper in front of you in case you would like to do any types of drawings.

Percipient:

(Pause) I'm just clearing my mind and trying to concentrate or think of where [Agent's] going to be at 2:15. (Pause) I'm getting some images and I—I can't discern whether they're really images from where she is or just, you know, prior images. (Pause) I'm seeing the—the arcade, I think it's called, next to the—of the Tribune building, Chicago Tribune building and I can't—I just cannot discern whether this is just—if it's part of the image of where she is or—or of an image that just popped into my mind from before. I'm seeing people walking by. There's a—a fountain. I don't know if it's a fountain, but there's a statue of some kind that leads to a point not particularly high. I'm get—I'm getting a statue face, on a statue or—that's high. That's fairly high, but I don't see a pedestal but I see the face of—on the statue. Uh, I'm getting the little turrets around the building and I can see the—the square or the—the, uh, area next to the building. It's not real deep but it's in a rectangular shape. I can see some people getting out of a taxi cab. There's a fair amount of movement around, of people going around or, you know, walking around. Uh—I think there must be a row of windows, thin, long windows but in a row of them going up quite high. (Pause) I'm seeing a—a heavy wooden door with—with a black bolt on it that—that go up in a—they're rounded at the top in a dome fashion I guess it would be called. I have a feeling of them opening the doors and looking in and it's dark inside. (Pause) My feeling at the moment is that it's a church, a building like a church. And I can see the—the pews. I feel basically darkness in the—there is some light but I feel basically kind of darkness in there and a quietness. (Pause) I'm again seeing the little turrets, very elaborate-looking little turrets, a whole series of 'em like across the entire top of the building and they—they—there's a straight line and then up to a triangular—triangle. I have a definite image of an angel-type of statue, marble, flowing robes. I can't make—maybe I shouldn't label it angel but I can—a statue with flowing robes. Although I'm—I'm seeing a halo, however.

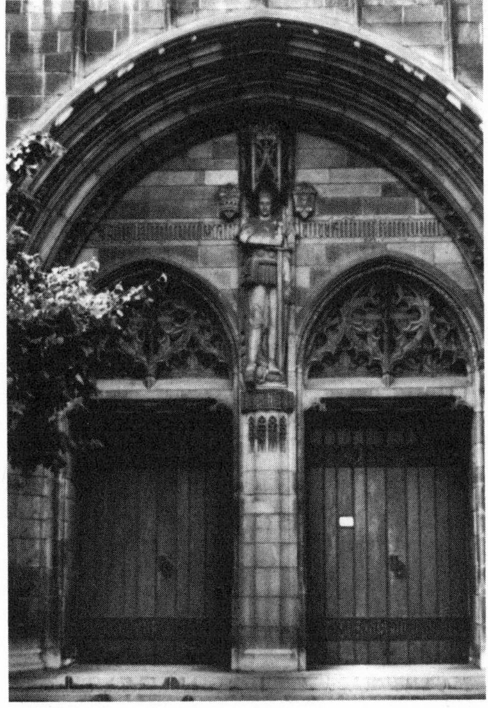

Fig. III-1a
Rockefeller Chapel, University of Chicago

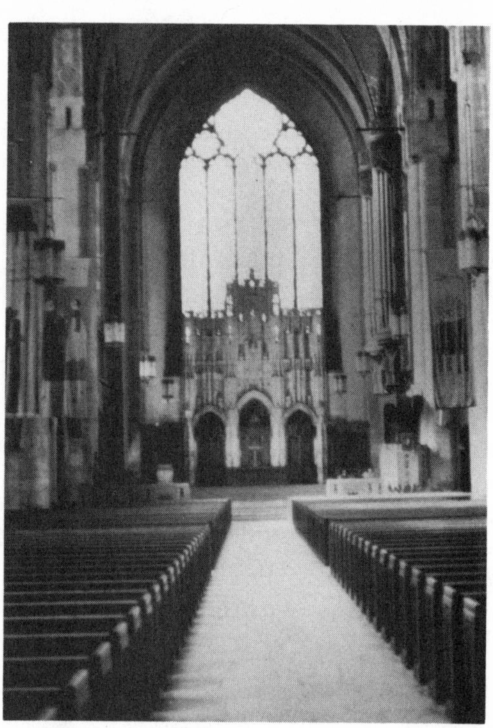

Fig. III-1b

(Pause) I see the door again and I see some stairs. (Pause) There are—there are some outside railings, I believe, that are on the stairs. And I don't think it's a particularly—I don't think it's—there're a whole lot of stairs. I think it's very high. (Pause) I'm getting some stained glass windows that are arch shape and they would look to be dark blue. I don't know whether there's any design on them or not but I—I have an impression of dark blue. (Pause) I want to say that I see an altar but—I—I'm having difficulty deciding whether I'm seeing the altar or whether I'm just thinking about it because I have the idea that it's a church. Whatever the architecture of it—the ornamentation on the building is—it's quite elaborate and it looks like there's a section on the top with the turrets and then below that there are some other kinds of designs, but more linear designs. Possibly scrolls within that linear design. I have the—a feeling of like a hard floor, like a marble floor. (Pause) And I have the image of which I just can't—I can't syphon out, you know, whether it's because I'm thinking of it, you know, because I think it's a church that I see a marble floor of like a pinkish neutral color and they're—I think they're in a square. It's

Fig. III-1c

fairly large squares. And I have a feeling of an aisle down the center. The—I believe there are other sets of doors, not—I can't tell whether they're just in the front or whether they're on the sides. I've more of a feeling there are doors on the sides but you don't necessarily see them as you look frontally at the building. I see the doors reappearing every once in a while, then again the arch doors with these black bolts on them, or black ornamentation. I see some trees, a very small area of trees of maybe just a few sparse trees, uh, be they in front or along the side or something. Not a park or anything but a feeling of some trees. (Pause) I see a fire hydrant. (Pause) I again have a vision of—of seeing the doors and then a maybe a ledged area or an area of the building that protrudes with some kind of a design and there may be even a couple of those before you get to the top part which is either triangular or rounded but it's—there are filigree work, little filigree turrets or something. On the inside I have a—I think it's on the inside, I have a feeling of—of the inside space or a space being arched. And within the building there is a sort of a continuation of arches but possibly they meet columns or something like that, but whatever the decoration is where walls join or separations or something there—looks like it's arched.

Interviewer:
Okay, [Percipient], it's time.

Percipient:
Okay.

The experimental procedure followed in this particular demonstration was similar to those routinely employed by the Mundelein group under the co-direction of one of the authors (BJD) from 1976 to 1979. During this period the group conducted a number of such trials confirming similar remote viewing results published a few years earlier by two physicists at SRI International.[2] These two bodies of data, along with a few other efforts within the parapsychological community, stimulated further activity in this form of psychic research and ignited considerable controversy with the critical community.

Despite the novel protocol of these experiments, the phenomena they address are hardly new. *Telepathy*—the perception of another person's thoughts and emotions —, *clairvoyance*—the acquisition of information from locations inaccessible by known sensory channels —, and *precognition*—the acquisition of information advanced in time—have been featured in shamanic and occult practice for eons, and in traditional psychical research over the past century. In personal and social contexts as well, the familiar "crystal ball" metaphor remains symbolic of man's innate desires to penetrate others' thoughts, to access hidden secrets, to foretell the future, or otherwise to reduce the randomness of his world.

Since the first SRI work and the Mundelein experiments, a number of other laboratories have mounted similar free-response studies and the composite data base on this genre of phenomena is now substantial. Although the various protocols have differed in some details, their basic approach has been generally similar to that just described: a percipient attempts to acquire information about a remote physical location or object by some anomalous means and to provide a description of the target in writing, on tape, or via sketches. This perception is then compared with the actual target scene by various subjective or analytical tests, on the basis of which some figure of merit is assigned. Methods for selection and definition of the targets also vary, some involving elaborate randomization techniques, others relying on more spontaneous selection of particular scenes or objects.

In most cases, as in the example above, a second participant, the agent, visits the target at a prescribed time, presumably to act either as a beacon for the percipient's effort or as an intermediary for the transmission of the information. The functional role of this agent, the efficacy of his possible strategies, and the importance of his relationship to the percipient are also interests of the research. In some of the studies, no human agent has been involved at all; rather, the target has been specified in terms of coordinates on a map or by other technical means.

Irrespective of the particulars, this type of experiment has become popular in contemporary parapsychological research because of its relatively high yield, low cost, and evident range of possible practical applications. It has also become a target of much critical commentary because of its high vulnerability to sloppy protocols and data-processing techniques. The most frequently cited criticisms include possibilities of inadvertent or deliberate sensory cuing of either participant, inadequately randomized target selection practices, shared conscious or subconscious preferences of the percipient and agent, and faulty statistical treatment. Given the inherently impressionistic character of the data, these may very well be legitimate concerns.

If such exercises are to qualify as credible research, therefore, rather than as interesting parlor games, sophisticated technical strategies need to be overlaid on their basically simple protocol. Beyond maintenance of adequate target randomization and experimental security, the most essential requirement is for standardized quantitative evaluation of the accuracy of the percipient responses. In virtually all of the early experiments cited above, some form of human judging was utilized where one or more persons not otherwise involved in the experiment compared the percipient's response with a number of alternative targets, including the actual one, and rank ordered these in terms of their quality of match to the perception. Alternatively, the percipients themselves were shown several targets and asked to rank them in order of agreement with their earlier perceptions. Simple statistical tests were then applied to the rank arrays. These methods share an assortment of methodological vulnerabilities, such as vagaries in the technical or perceptual capabilities of the judges; their subjective biases toward individual experiments or toward the topic in general; their inability to compare more than a few alternatives in any given case; and even possible psychic input on their own part. Beyond these problems, the methods are statistically very inefficient, in that the entire content of a given perception is reduced to a single datum, despite its potential richness of explicit and implicit information.

For these reasons, the remote perception portion of our PEAR program has addressed itself as much to the development and application of methods for detailed analytical evaluation of experimental results as to the accumulation of data *per se*, and it is to this analytical effort that this section is primarily addressed. We shall begin, however, with a concise statement of our own protocol, a representative selection of perception results, and a brief summary of their characteristic features, on the basis of which the analytical methodology can be more tangibly described.

2

PEAR PERCEPTIONS

In its present form, our formal remote perception protocol requires the percipient to generate a written description of an unknown geographical target where the agent is, was, or will be stationed at some prescribed time, and then to fill out a check sheet of questions on which subsequent analytical judging schemes are based. Percipients select their own time and place to generate and record their target impressions and employ their own varied strategies for the task. The styles of their written transcripts also vary widely, from cryptic, analytic statements on one extreme, to lengthy, detailed descriptions on the other.

The agent typically spends ten to fifteen minutes at the target, beginning at the assigned time, and records his impressions of it in prose and via the same checklist. Whenever possible, photographs are also taken of the scene to corroborate the description and for possible future reference. Again, there are no constraints on the subjective strategies employed by the agent, which may range from casual indifference to intense concentration on the scene, on the percipient, or on both.

Target locations are determined in either an instructed or a volitional mode. In the instructed mode, the target is randomly selected before each trial from a large pool of targets previously prepared by an individual not otherwise involved in the experiment and maintained so that no potential percipient or agent has access to it. In the volitional mode, the target is selected spontaneously by the agent at the time specified for the trial. This latter protocol is most commonly employed when the agent is traveling on an itinerary unknown to the percipient, in a region for which no prepared target pool exists.

However the target is selected, most of the perceptions are performed precognitively. That is, they are generated and recorded before the target is visited by the agent, indeed before the target is even selected, and no sensory communication between percipient and agent is permitted until all descriptions, photographs, and check sheets are recorded and filed. A smaller number of trials have been executed retrocognitively, where perceptions are generated after the target visitation, but before the percipient has any opportunity to obtain information about the target by ordinary means. Since more than visual imagery seems operative in any of these protocols, we prefer the more generic title Remote Perception to the original Remote Viewing designation; and since most of our data are obtained in the precognitive format, our acronym for this portion of the program is PRP—Precognitive Remote Perception.

At this time, the PEAR files of PRP data contain 334 formal trials obtained by some 40 percipients under the various protocols outlined above. In addition, we retain a file of all informal or exploratory trials that, for one reason or another, do not conform to all of the requirements of the formal protocols but nonetheless may embody instructive information. A full compendium of all trials, including numerous examples of target photographs, associated perception transcripts and check sheets, and extensive tabular results derived from the analytical judging methods, is available in a technical report.[3] Here we shall only illustrate by a few selected examples some of the salient features of this type of data and its evaluation.

Figures III-2–7 display photographs of various PRP targets with notations of their location, time of visitation by the agent, and time and location of the perception effort, followed by the percipient's verbal transcript. Note the variations in style, detail, precision, and emphasis prevailing in these perceptions that must somehow be accommodated by any quantitative analysis technique. From examination of our full data base, a number of generic characteristics can be noted that may be indicative of the nature of the process. For example, the perceptions range in accuracy from virtually photographic precision, through varying degrees of correspondence with the components and ambience of the scene, to total irrelevance. In some cases, specific features are properly identified but placed in incorrect relationships or environment; in other cases, the overall ambience is accurately perceived, but some of the elements are incorrect. Details that are central to the scene, or to the agent's sense of it, may be totally ignored by the percipient while minor aspects are escalated to importance. Major geometrical distortions are common, and the more aesthetic aspects seem to be perceived more accurately than the analytical details. In some cases, there is a progression in the transcript from accurate to inaccurate material, or vice versa. Even these generalizations are susceptible to individual interpretation and are shaky ground on which to base claims of anomaly or attempts to correlate and comprehend the phenomena. The key question thus becomes whether more systematic quantitative assessment techniques can be developed to render the impressionistic data into fungible form.

Fig. III-2

Northwest Railroad Station, Glencoe, Illinois

Perception (in part):

"I'm seeing an overpass right now, an overpass, probably over the expressway. I see it banked in the grass along the side, sloping up. I see a train station. I see the train station, one of the commuter train stations that's on the expressway—the white cement of them and the silver railings. I see a train coming. . . . There are a lot of buildings around, but I think they're houses. . . . I keep seeing the traffic going by speedily, very rapidly, a lot of cars. . . . See just the front end of the train station. See a little bit within it. . . . I think there are wooden planks on the floor. . . . I see or hear the clicking of feet or shoes on the wooden floor. . . . There are posters or something up, some kind of advertisements or posters on the wall in the train station. I see the benches. Getting the image of a sign, but I think it's probably the sign of what station it is. It's about eight or ten letters in the word. Maybe something like Clydeburn or Clayburn. Have the impression of this wooden floor being somewhat littered, just sort of dirty. I see the tracks. No train on the tracks right now, empty tracks. . . ."

Distance: 5 miles
Time: 35 minutes, precognitive
Normalized Score: .850
z-Score: 2.862
Probability: .002

Fig. III-3

Radio Telescope, Kitt Peak, Arizona

Perception (in full):

"Rather strange yet persistent image of [agent] inside a large bowl—a hemispheric indentation in the ground of some smooth man-made material like concrete or cement. No color. Possibly covered with a glass dome. Unusual sense of inside/outside simultaneity. That's all. It's a large bowl. (If it was full of soup [the agent] would be the size of a large dumpling!)"

Distance:	2200 miles
Time:	45 minutes, precognitive
Normalized Score:	.731
z-Score:	1.883
Probability:	.030

Fig. III-4

Terrace, Milan, Italy

Perception (in full):

"Agent sitting on a small porch or balcony surrounded by a stone or brick wall, maybe waist high, talking to another person. Feet up, very relaxed physically, lots of energy. View from balcony looks down (maybe one story high) onto a rather narrow street, somewhat curved or winding, lined with buildings. Colorful scene, street sounds, some vehicular traffic—not heavy—pedestrians. A few trees, not necessarily on street—maybe on balcony—flowers in pots. Sunny warm day."

Distance: 4000 miles
Time: 132.5 hours, precognitive
Normalized Score: .706
z-Score: 1.673
Probability: .047

Fig. III-5

Ruins of Urquardt Castle, Loch Ness, Scotland

Perception (in full):

"rocks; with uneven holes; also smoothness
height
ocean—dark, dark blue
  whitecaps
  waves; booming against rocks?
on mountain or high rocks overlooking water
dark green in distance
gulls flying?
  pelican on a post
sand
a lighthouse?—tall structure—round with conical roof
  high windows or window space with path leading up to it
  or larger structure similar to a castle
  [Here there is a sketch of a castle abutment on the transcript.]
a black dog?
  with longish curly hair
old, unused feeling; fallen apart?
musty or dank
moss or grass growing in walls
drawbridge?—wood
ocean smells
[Here two sketches show the rocks and castle separated by water from the dark green in the distance.]
flowers?
snow ice—capping a mountain?
high large, cavernous hall—castle"

Distance: 3500 miles
Time: 14.5 hours, retrocognitive
Normalized Score: .769
z-Score: 2.194
Probability: .014

Fig. III-6

Tretiakovskaia Gallereia, Moscow, U.S.S.R.

Perception (in part):

"Have the sensation of being in a very quiet, somber, subdued sort of atmosphere. Something like the feeling you get in an old cathedral or cemetery. Possibly an old castle or something like that. Maybe a fortress. . . . Any color impressions I get are the same—grays, browns, dark subdued colors. I feel an oldness. . . . I'm thinking of a large church or something, or a castle. Some kind of building. It seems to be quite large. Sensation of sounds echoing, subdued colors. . . . I see several, maybe two to four round balls that seem to be on top of something. Maybe it's some kind of decoration. Like on top of something that's of a generally square shape. Almost like a square column with a ball on top. I have a very clear picture suddenly of an old building. It's quite large. There are windows with like arches. They may not be exactly arched, the arches come to a point on top almost. Very impressive. It's a light gray color, very ornate. It comes to a point of some sort, but it's not a regular point. . . . Like where it should be round on top it comes to a point. I'm not sure if it's windows or the shape of the building itself. . . . Great big double doors. . . . Just saw those square pillars with the balls on top again. They seem to be almost like an entrance way, one on either side. . . ."

Distance: 6000 miles
Time: 21.5 hours, precognitive
Normalized Score: .664
z-Score: 1.323
Probability: .093

Fig. III-7

Danube River, Bratislava, Czechoslovakia

Perception (in part):

"I have the feeling that [the agent] is somewhere near water. I seem to have the sensation of a very large expanse of water. There might be boats. Several vertical lines, sort of like poles. They're narrow, not heavy. Maybe lamp posts or flagpoles. Some kind of circular shape. Almost like a merry-go-round or a gazebo. A large round thing. It's round on its side, like a disc, it's like a round thing flat on the ground, but it seems to have height as well. Maybe with poles. Could possibly come to a point on top. Seeing vertical lines again. Seems to be a strong impression, these vertical lines. No idea what they could be. . . . A definite sensation of being outside rather than in. Water again. . . . To one side of where [the agent] is I get the feeling there's a kind of small building. Could be a shed. . . . Predominant colors seem to be blue and green. . . . Water again. Some very quick impression of a fence, a low fence. . . . Steps. I don't know where they're leading to. . . . The steps sort of lead up to like a path or walkway. Like a boardwalk. And there's a fence along it. There's people walking along it, and there's vertical lines along that walkway. . . ."

Distance: 5600 miles
Time: 23.5 hours, precognitive
Normalized Score: .884
z-Score: 3.145
Probability: .001

3

ANALYTICAL ASSESSMENT

Over recent years, a variety of computer-facilitated techniques in artificial intelligence and pattern recognition have been developed for extraction of specific quantitative details from various complex patterns of information. Application of similar strategies to assessment of remote perception transcripts may eventually also prove useful, but given the highly subjective and holistic character of these data and the primitive nature of our present comprehension of the process, a more rudimentary approach seems preferable at this time. Our tactic is to establish a code or "alphabet" of descriptive queries that can be addressed to all targets and perceptions. In their simplest version, these descriptors, thirty in number, are posed in binary form, to be answered yes or no, and range from factual discriminations, such as whether the scene is indoors or outdoors, whether trees are present, or whether there are automobiles, to more subjective aspects, such as whether the ambience is noisy or quiet, confined or expansive, hectic or tranquil. The full list of binary descriptor questions are presented to the participants in the following order:

1) Is any significant part of the perceived scene indoors?

2) Is the scene predominantly dark; for example, poorly lighted indoors or nighttime outside?

3) Does any significant part of the scene involve perception of height or depth, such as looking up at a tower, tall

building, mountain, vaulted ceiling, or unusually tall trees, or looking down into a valley or down from any elevated position?

4) From the agent's perspective, is the scene well-bounded, such as the interior of a room, a stadium, or a courtyard?

5) Is any significant part of the scene oppressively confined?

6) Is any significant part of the scene hectic, chaotic, congested, or cluttered?

7) Is the scene predominantly colorful, characterized by a profusion of color, or are there outstanding brightly colored objects prominent, such as flowers or stained-glass windows?

8) Are any signs, billboards, posters, or pictorial representations prominent in the scene?

9) Is there any significant movement or motion integral to the scene, such as a stream of moving vehicles, walking or running people, or blowing objects?

10) Is there any explicit and significant sound, such as an auto horn, voices, bird calls, or surf noises?

11) Are any people or figures of people significant in the scene, other than the agent or those implicit in buildings or vehicles?

12) Are any animals, birds, fish, or major insects, or figures of these, significant in the scene?

13) Does a single major object or structure dominate the scene?

14) Is the central focus of the scene predominantly natural, that is, not man-made?

15) Is the immediately surrounding environment of the scene predominantly natural, that is, not man-made?

16) Are any monuments, sculptures, or major ornaments prominent in the scene?

17) Are explicit geometric shapes—for example, triangles, circles, or portions of circles (such as arches), or spheres or portions of spheres (but excluding normal rectangular buildings, doors, windows, and so forth)—significant in the scene?

18) Are there any posts, poles, or similar thin objects, such as columns, lamp posts, or smokestacks (excluding trees)?

19) Are doors, gates, or entrances significant in the scene (excluding vehicles)?

20) Are windows or glass significant in the scene (excluding vehicles)?

21) Are any fences, gates, railings, dividers, or scaffolding prominent in the scene?

22) Are steps or stairs prominent (excluding curbs)?

23) Is there regular repetition of some object or shape, for example, a lot full of cars, marina with boats, or row of arches?

24) Are there any planes, boats, or trains, or figures thereof, apparent in the scene, moving or stationary?

25) Is there any other major equipment in the scene, such as tractors, carts, or gasoline pumps?

26) Are there any autos, buses, trucks, bikes, or motorcycles, or figures thereof, prominent in the scene, moving or stationary (excluding agent's car)?

27) Does grass, moss, or similar ground cover compose a significant portion of the surface?

28) Does any central part of the scene contain a road, street, path, bridge, tunnel, railroad tracks, or hallway?

29) Is water a significant part of the scene?

30) Are trees, bushes, or major potted plants apparent in the scene?

In more elaborate formats, the descriptors are posed in ternary or quaternary notation where the percipients and agents are allowed three or four response options that permit them to neutralize or reject a given question if they feel that it is irrelevant or uncertain, or to indicate the relative importance or centrality of a particular feature. For example, in ternary form, question 9 regarding motion in the scene may be phrased: Is motion of any kind a) present; b) unclear or uncertain; or c) absent? In quaternary form, this same descriptor may be written: Is motion of any kind a) dominant or central; b) clearly present, but not dominant; c) unclear, uncertain, or irrelevant; or d) clearly absent.

Whatever the particular form of the descriptors, these methods render the targets and perceptions into arrays of digits that allow computerized scoring algorithms to be applied in totally objective, very rapid, and very prolific fashions. Unlike a human judge, who can reasonably rank only a few alternative targets, the computer can score and rank any digitized perception against all of the hundreds of targets in the pool, if desired. Indeed, the most powerful aspect of this technique is that by scoring every perception against every possible target except the actual one, a large array of deliberately mismatched scores compounds that may serve as an empirical chance distribution, in the sense that it may be used as a reference distribution for statistical analysis of any single perception, or group of perceptions, scored against its proper target (Fig. III-8).

A number of scoring recipes have been applied to the various descriptor response arrays provided by the percipients and agents. The simplest of the binary methods merely counts the number of descriptors on which the percipient and agent agree, but this approach embodies a serious flaw: the natural probability of any given descriptor throughout the entire target pool being yes/no is seldom precisely .50/.50. For example, more scenes tend to be outdoors than indoors, more tend to have people present, fewer have animals, etc. Therefore, a

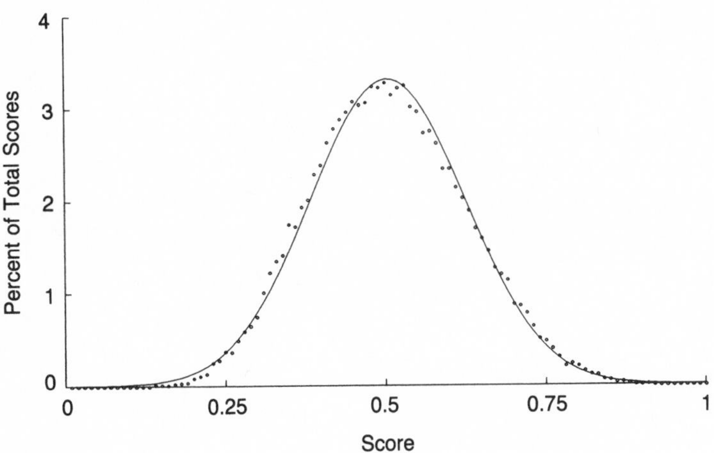

Fig. III-8 *PRP Empirical Chance Distribution: 42,000 Mismatched Scores with Gaussian Fit*

percipient who correctly identifies a less likely descriptor should receive greater credit than for correct response to a more likely one. This flaw may be rectified by assignment of a weighting factor to each descriptor based on its *a priori* probability throughout the target pool, as assessed by the computer from the full target data. Using these weighting factors, scores may then be compounded by a formula similar to that used in gambling payouts. Specifically, if the *a priori* probability of a "yes" response is α, the score given a correct "yes" response is $1/\alpha$ and that given a correct "no" response is $1/(1\text{-}\alpha)$.

To illustrate this analytical scoring method, Table III-1 shows the perception and target binary response digits for the Rockefeller Chapel trial with which this section began. Also listed are the *a priori* descriptor probabilities for the relevant target pool. The final column tabulates the score increments provided by each descriptor, the total of which constitutes the absolute score of the trial, in this case, 46.822. This absolute score is then divided by the maximum score that could have been obtained if all of the responses had agreed, 56.297, yielding a normalized score of 0.832. The statistical merit of this normalized score is established by comparing it with the distribution of similarly normalized empirical chance scores computed from the array of perception/target mismatches described above. For this target pool and this binary scoring method, that empirical chance distribution has a normalized mean of 0.504 and a standard deviation of 0.121; hence this Chapel trial score stands 2.7 standard deviations to the high side of the mean, a result unlikely by chance to the order of .003. Scores computed by this same method are also included in the examples shown in Figs. III-2–7, and indicated on the empirical chance distribution reproduced in Fig. III-9.

The analytical results of our entire PRP data base are illustrated in Fig. III-10, where the distribution of the 334 perception scores (solid line) is compared to the empirical chance distribution compounded from 42,000 mismatch scores (dashed line). The distortion of the proper score distribution to the high side is indicative of the extra-chance information

TABLE III-1

Descriptor Contributions
Rockefeller Chapel

Descriptor	*A Priori* Probability (α)	Target	Perception	Score Increment
1. Indoors	.397	1	1	2.519
2. Dark	.215	1	1	4.651
3. Height	.486	1	1	2.058
4. Bounded	.589	1	1	1.698
5. Confined	.154	0	0	1.182
6. Hectic	.229	0	0	1.297
7. Color	.537	1	1	1.862
8. Signs	.458	1	0	0
9. Motion	.621	1	1	1.610
10. Sound	.715	1	0	0
11. People	.710	1	1	1.408
12. Animals	.248	0	0	1.330
13. Single Object	.439	1	1	2.278
14. Natural Focus	.154	0	0	1.182
15. Natural Environment	.360	0	0	1.563
16. Monuments	.290	1	1	3.448
17. Shapes	.743	1	1	1.346
18. Poles	.650	1	0	0
19. Doors	.547	1	1	1.828
20. Glass/Windows	.706	1	1	1.416
21. Fences	.495	0	1	0
22. Stairs	.304	1	1	3.289
23. Same	.621	1	1	1.610
24. Planes	.206	0	0	1.259
25. Equipment	.266	0	0	1.362
26. Vehicles	.477	1	1	2.096
27. Grass	.421	1	0	0
28. Roads	.612	1	1	1.634
29. Water	.257	0	0	1.346
30. Trees	.645	1	1	1.550
			Total Score	46.822

Maximum Possible Score for Target:	56.297
Normalized Score for Perception:	46.822 ÷ 56.297 = 0.832
z-Score for Trial:	2.716
Probability by Chance:	.003

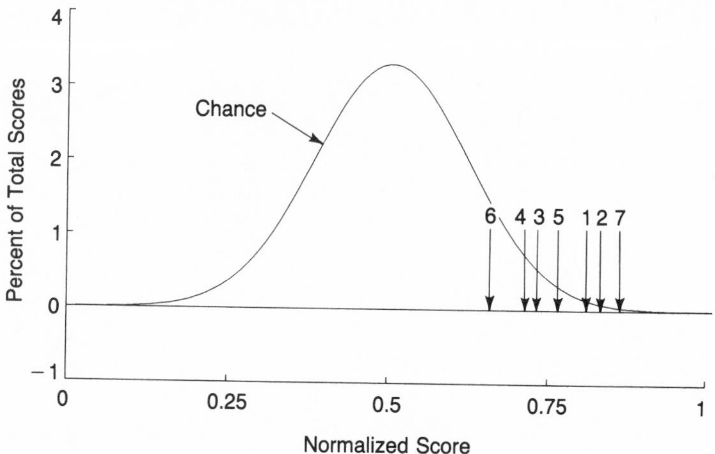

Fig. III-9 *PRP: Scores of Illustrated Trials (Figures III-1 to III-7) Compared with Empirical Chance Distribution*

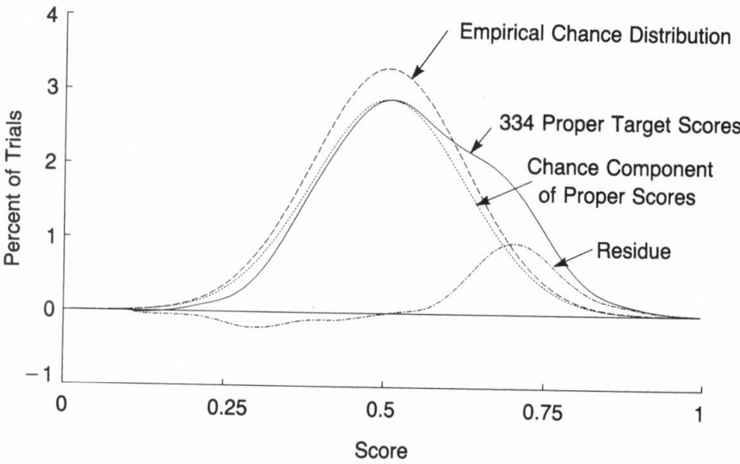

Fig. III-10 *PRP Score Distributions*

acquisition over the total data base, and can be quantitatively evaluated by a number of methods. For example, if we subtract from the proper score distribution the largest component of it that can be attributed to chance, namely the largest subset of it that conforms to the empirical chance distribution parameters (dotted line), then the residue distribution (dot-dashed line) should represent the extra-chance information component. For these data, about 15% of the integrated results reside in this extra-chance residue. Detailed statistical calculation yields a probability for acquiring this much information by chance of about two parts in 10^{11}. Table III-2 displays a numerical summary of these same data.

All told, some twenty binary, ternary, and quaternary scoring methods have been developed for application to our PRP data base. Each has particular strengths and weaknesses in rendering impressionistic information into quantitative analytical form, and each embodies certain trade-offs of scope and detail versus ease of response and calculation. Nonetheless, the results of these diverse methods are reassuringly consistent in their overall assessment of the quantity and character of the information acquired via the PRP process. In particular, they leave little doubt that substantial pragmatic information has been obtained by many percipients, using a variety of personal strategies, addressing a broad variety of targets. We are thus justified in turning to the dependence of such phenomena on the controllable technical, physical, and personal parameters of the experiments.

TABLE III-2
Precognitive Remote Perception
Data Summary

	All Data	Instructed	Volitional
Number of Trials	334	125	209
Mean Score	.5480	.5662	.5371
Chance Mean	.5042	.5042	.5042
Chance Std. Dev.	.1207	.1207	.1207
z-Score	6.622	5.738	3.934
Probability	1.8×10^{-11}	4.8×10^{-9}	4.2×10^{-5}
No. Trials $p < .05$	47 (8)*	20 (4)	27 (4)
No. Trials $p < .5$ [%]	208 [62%]	85 [68%]	123 [59%]

* Numbers in ( ) indicate trials with negative z-scores, $p < .05$

4

PENETRATION PARAMETERS

Understanding of the remote perception anomaly may be aided by empirical demonstration of its sensitivity to particular physical or psychological parameters that can be systematically varied over a substantial number of trials. At this stage we have adequate data to comment on just three such variables: the distance between the percipient and the target; the time interval between the perception effort and the actual visitation of the target by the agent; and the relative effectiveness of the volitional and instructed protocols. On other potential correlations, we have only anecdotal indications.

The distance between the percipient and the target suggests itself as a primary discriminator among the various conceivable means of such information acquisition. If, as sometimes postulated, some form of physical wave propagation, such as electromagnetic waves or geophysical waves, were involved, an inverse dependence of the fidelity of the information on that distance would be expected. However, as illustrated in Fig. III-11, within the statistical variance of these data and the analytical treatments employed, no such significant dependence is found. Up to intercontinental distances of several thousand miles, there is no discernible deterioration in the signal-to-noise ratio.

Such physical wave models are further contested by the similar absence of any evident dependence of perception accuracy on the time interval between perception effort and target visitation. Figure III-12 shows the same scores plotted as a function of that time displacement. Positive time intervals denote perceptions dictated precognitively by the indicated number of hours. Negative times denote data acquired retrocognitively, where the percipient dictated his perception after the target had been visited. Again, over the interval range of

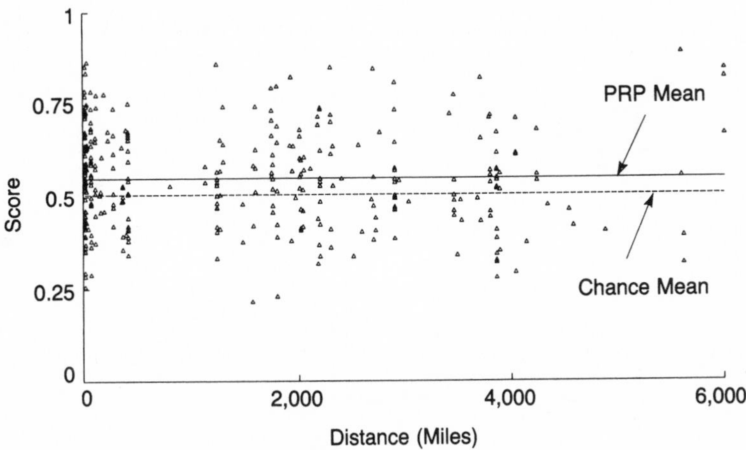

Fig. III-11 *PRP Scores as a Function of Distance*

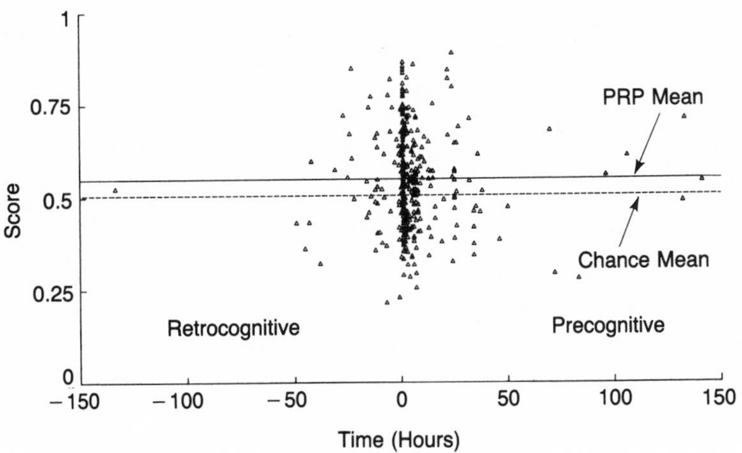

Fig. III-12 *PRP Scores as a Function of Time Interval*

several days covered by these experiments, there is no statistically significant dependence on these time intervals.

Comparison of results for data acquired following the volitional protocol, where the target is selected spontaneously by the agent at the time assigned for visitation, with those under the instructed protocol, where the agent is directed to a specific target randomly preselected by a third party, might also help to discriminate the mechanisms involved. For the volitional mode, one could speculate that the agent's selection of the target might be influenced by some sense of preference of the percipient, acquired either telepathically or simply via some knowledge of the latter's predisposition to particular types of objects or scenes. In the instructed mode, however, such biased target selection is largely precluded. Thus, if sympathetic target selection is a major factor, one would expect better scores under the volitional protocol. As shown in Table III-2, such is not the case. On a statistical basis, there is very little to choose between the pattern of volitional and instructed scores.

This should not be taken to imply that psychological resonance between the percipient and agent is an unimportant parameter of the experiment. To the contrary, there is a substantial body of anecdotal evidence from our work and that of others that various forms of emotional or physiological bonds between the participants can facilitate the process. On the other hand, we have occasional successful trials using virtual strangers and, in a few cases, persons quite skeptical of the process. Other laboratories have reported success using no agent at all. Unfortunately, this will not be a simple issue to resolve, since large data bases for each percipient/agent pair will be required for legitimate statistical assessment and since sufficiently incisive psychological indicators useful in quantifying such correlations have not yet been identified.

Similar demurral must be made regarding the role of individual percipient and agent strategies, attitudes, and personalities. These range so widely, and frequently vary so substantially for given participants from trial to trial, that the data accumulation task preceding any legitimate attempts at correlation is formidable. What may be peripherally relevant, however, is the observation that the overall remote perception case does not rest predominantly on a few outstandingly accurate perceptions immersed in a sea of chance results; rather, the yield compounds incrementally over a large number of trials having individually marginal achievements. For example, about 14% of the 334 trials are individually significant by the .05 criterion; however, over 62% score above the chance mean. In this sense, the phenomenon is similar to the REG and RMC effects described earlier. Indeed, a cumulative deviation graph compounding all formal PRP scores (Fig. III-13) resembles the individual and collective operator signatures of the man/machine experiments.

Finally, one might turn to the range of characteristics of the targets themselves to search for patterns of correlation of success with their details or ambience. Here again, the burden of data collection prior to comprehensive assessment is immense, but we have certain indirect indications. As mentioned earlier, the various descriptors have been selected to provide a balanced, incisive alphabet of feature discriminators, ranging from objective and detailed aspects to quite subjective and holistic ones. The effectiveness of these descriptors in building

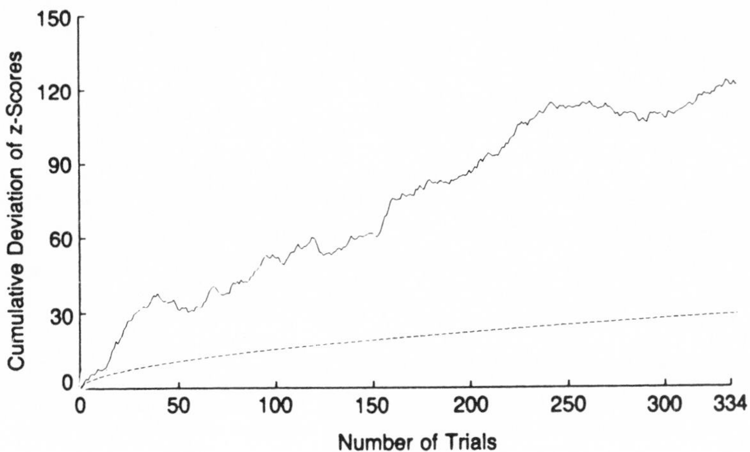

Fig. III-13 *PRP Cumulative Deviation of z-Scores*

the various trial scores is readily assessed by subroutines in the scoring algorithms. By such tests, it is found that the entire group comprises a remarkably uniform set in terms of their effectiveness in quantifying information across the full range of target types. In fact, the efficacy of individual descriptors, integrated across the total target pool, differs only marginally from chance expectation. None is exceptionally effective; none is seriously deficient. Only in their systematic combinations over large numbers of trials do they convey the anomalous information. The implication is that specific target features, as assessed by these descriptors, are not parameters of primary importance.

5

ANECDOTAL ADDENDA

It is critical for anomalies research in general, and for PRP studies in particular, to guard against excessive reaction to, or dependence on, extraordinary incidental results. For this reason, the PEAR program works only with very large quantitative data bases acquired under systematic and rigorous protocols. However, this caution need not exclude the consideration of certain anecdotal effects as possible guideposts to more deliberate experimental explorations or to hypothetical propositions that can be systematically tested. The PRP experiments are a particularly rich source of such intriguing anecdotes, and some additional insight into the underlying process has been gleaned from them.

In one exploratory trial, for example, a percipient in Princeton, New Jersey, provided a reasonably accurate precognitive description of a street scene in Paris where the agent was immersed in a typical ambience of human bustle, traffic flow, and shop-front paraphernalia. He also perceived an image of a medieval knight in full armor somehow involved in this pattern, and acknowledged this unlikely element in his free-response transcript. The agent had no corresponding feature in his target description but, after reading the perception, recalled that he had been standing in front of a government building ornamented with statuary of historical military figures attired in period dress. One of these closely matched the percipient's description.

In a second instance, an agent selected as target the components of the Saturn moon rocket at the NASA Space Center in Houston, Texas. The Princeton percipient described an indoor scene with the agent playing on the floor with a group of puppies—a complete "miss," both impressionistically and analytically. Later that evening, before learning any details of the perception, the agent visited a friend's home where he played at length with a litter of newborn pups, one of which he was prompted to purchase.

In a somewhat similar case, an agent en route to a target of a gambling casino in Nevada stopped at a service station and amused himself for a while by attempting to ride a collapsible bicycle. The perception transcript, dictated in Chicago, included nothing relevant to a casino but featured an accurate description of the agent's clumsy attempt to ride the bike.

Each of these examples, along with several others in the files, could be interpreted in terms of some spatial or temporal displacement in the percipient's identification of the target that yielded correct information about an adjacent event or location but had to be technically discounted by the formal protocol and analysis. Recall that in the lengthy perception of the Chicago chapel, the percipient began by first correctly identifying the Tribune building, even though this was not a proper portion of the spatially or temporally circumscribed target. In other cases, we find different kinds of displacement effects that can actually improve the scores. For instance, an agent standing on an Idaho farm was concentrating on a herd of cows when his attention was drawn to an irrigation ditch several yards down the road. This impetus was sufficiently strong that he included the ditch in one of the target photographs and acknowledged it in the descriptor response. The perception, acquired precognitively from New Jersey, did not mention cows at all, but did specify farm buildings, fields, and an irrigation ditch.

In another example, the agent had chosen as target an indoor seminar room in a Montana lodge. Midway through the assigned period, his mind inadvertently wandered to a scene just outside the building that involved a low stone wall, many trees, and a picnic table, all of which were then noted in the target description. The New Jersey perception ignored the interior scene but was virtually photographic in its identification of the details and ambience of this outside portion of the target.

The message we derive from the full assortment of such anecdotal effects reinforces the essential dilemma of this research: by its nature, the remote perception process is inherently impressionistic in character, yet to render it scientifically credible and pragmatically dependable, it must be reduced to analytical and quantitative terms. Thus we find ourselves fishing in a metaphysical sea with a scientific net far better matched to other purposes. Inevitably, much of the anomalous information we seek will slip through this net, leaving us only skeletal evidence to retrieve, but it is on that alone that we can base systematic analysis and scientific claim.

6

PRACTICAL POSSIBILITIES

When all present evidence is digested, these salient features of the PRP phenomenon can be defended:

1) Using the experimental protocols and analytical scoring methods outlined above, individual percipients can acquire statistically significant information about spatially and temporally remote target locations by means currently inexplicable by known physical mechanisms.

2) This acquisition of anomalous information compounds across the data base mainly by small increments beyond chance in a large number of trials, supplemented by a lesser number of highly successful efforts.

3) In any given trial, the major components of information may be localized in a few specific descriptors, but across the full data base each descriptor contributes comparable amounts of information.

4) For this data base and these methods of analysis, no evidence is found to suggest deterioration of the acquired information with increasing distance between percipient and target, up to several thousand miles.

5) Similarly, no evidence is found to suggest deterioration of the information with the degree of precognition, or retrocognition, up to several days.

6) While insufficient data and indicators exist to assess the role of the percipient-agent relationship or of their individual

strategies, personalities, or attitudes in any critical detail, anecdotal experience suggests that the process is intrinsically imprecise and bidirectional, and hence subject to distortions and displacement effects.

Even in this limited context, the PRP data base constitutes another substantial anomaly in the path of established physical science, supplementary, and possibly complementary, to the man/machine anomalies discussed in Section II. In this case, however, as often occurs in the evolution of science, numerous empirical applications and ramifications of the basic phenomenon have long preceded its rigorous demonstration and theoretical comprehension. Historically, human desire to access hidden information or to forecast future events has manifested itself in multivaried pragmatic applications of similar psychic processes, and these strategies persist in modern societies. Even in the most technological cultures, the utility of specific remote perception protocols and data-analysis techniques for anomalous acquisition of information is currently being explored to varying degrees in such areas as national security, law enforcement, missing-persons searches, natural-resource location, archaeological prospecting, medical practice, business forecasting, and personal enhancement, and a variety of training programs for cultivation of such skills are commonly advertised.

Although we shall return to such possible applications, our central concern is the incorporation of controlled experimental findings into a comprehensive model of the role of consciousness in the establishment and perception of physical reality, and for this task the specific characteristics of the phenomena summarized above should provide valuable hints. Certainly, if such a model can render more credible Paracelsus's sweeping assurance that headed this section, or help to reestablish the theoretical foot of science on better ground for further advance into this metaphysical domain, the implications for personal and collective human experience will be great.

SECTION FOUR

The Waves of Consciousness

O Nature, and O soul of man! how far beyond all utterance are your linked analogies! not the smallest atom stirs or lives in matter, but has its cunning duplicate in mind.

—HERMAN MELVILLE, *Moby Dick*[1]

Moby Dick, by Rockwell Kent

1

MODEL MAZE

Our experimental expeditions into the realm of consciousness-related phenomena have so far uncovered two species of anomalies that are starkly inconsistent with established physical theory. In one case, the tangible output of a variety of information-processing devices has been shown to be correlated with the stated intentions of human operators in statistically replicable and individually characteristic fashions. In the other, substantive information about geographical targets inaccessible by any known sensory channel has been acquired by remote percipients, with a degree of fidelity that appears to be statistically insensitive to the intervening space or time. Although much more data will be required to specify these and similar anomalies with any completeness, it is not too soon to begin serious contemplation of their theoretical accommodation, in terms that enable the scientific two-step to proceed.

> **"The first step in intuition is to understand the symbolical meaning of different things, and the next is to express them symbolically."**
> HAZRAT INAYAT KHAN

For this purpose, it is first necessary to establish some pattern of conceptualization and nomenclature that lends itself to representation and discussion of the observed effects. Next, some mathematical formalism capable of quantifying the proposed mechanisms and pertinent parameters needs to be devised. These concepts and formalism must then be deployed in some schema of logical inference and analytical computation that confirms established evidence, suggests new experiments, and makes testable predictions about their outcomes. By successive iterations of such theoretical models with progressively more incisive experiments, the ultimate goal of fundamental comprehension of the phenomena may systematically be approached.

These are not new aspirations. The literature of psychic research over the past century abounds with a smorgasbord of attempts to model all manner of paranormal effects, with very limited success. Without attempting any compendium of this body of effort, we simply observe that it tends to sort into four broad categories:

1) a skeptical genre that presumes that the effects are totally illusory and then attempts to explain them as artifacts of poor experimentation or data processing, or as chance results of complex random processes;
2) extensions of prevailing psychological models of memory, learning, or motivation or those that endeavor to correlate operator performance with various indicators of personality, attitude, or strategy;
3) theories that postulate established physical mechanisms, such as electromagnetic waves or geophysical processes, to convey information or influence to or from the human neurophysiology;
4) models that entail major modifications, extensions, or generalizations of established scientific concepts and formalisms to accommodate the effects at issue.

The skeptical explanations constitute an interesting and instructive body of effort, some portion of which has indeed proven correct and valuable in disqualifying invalid evidence that could have misled subsequent research, and another portion of which has illustrated comparable foibles on the part of the skeptics themselves. While these interpretations have provided valuable cautions in the design and operation of our experiments, we believe that the sophistication of the equipment, the discrimination of the tripolar protocols, the large data bases, and the redundancy of the automated data handling are adequate to preclude technical or statistical artifacts or operator fraud. Beyond this, the specificity and replicability of individual operator signatures and the degrees of statistical significance argue forcefully against chance occurrence.

The psychological models also have proven helpful in developing our experiments and laboratory environment, and in contemplating some of the empirical results, but are of limited value beyond that, for two reasons. First, the focus of all of these models is primarily on the human operator as "subject," whereas our experimental work has preferentially addressed the physical and technical parameters of the interaction. Second, the subjective concepts and nomenclature employed in these models lend themselves to little quantification, and then only in terms of psychological parameters that are difficult to assess.

The models that attempt to explain the phenomena in terms of known physical and neurophysiological processes furnish a more amenable conceptual and mathematical vocabulary for discussion of our results and clearly should be carefully considered before any fundamental modifications of the theoretical structure are contemplated. The most familiar approach of this class, and historically the first proposed in any systematic terms, presumes an electromagnetic basis of interaction. Over more or less the same period in which controlled psychic research data were first beginning to compound, contemporary science and technology were providing the world with wireless radio communication. The extraordinary capabilities of this new mode of information transfer probably conditioned the imagery of some of the researchers attempting to understand various psychic effects, and frequent reference to "transmitters," "receivers," "antennas," "tuning," and "resonance" can be found in their writings.

Attempts to apply electromagnetic models in quantitative analytical detail appeared somewhat later, both in the Soviet literature and in the West, and have persisted in various formats to the present. Many of these models have focused on extremely low frequency (ELF) electromagnetic effects, of the order of 1 to 50 hertz (cycles per second), presumably because of the similar frequency range of many neurophysiological processes that might conceivably be involved in the human reception or radiation of information by this means.

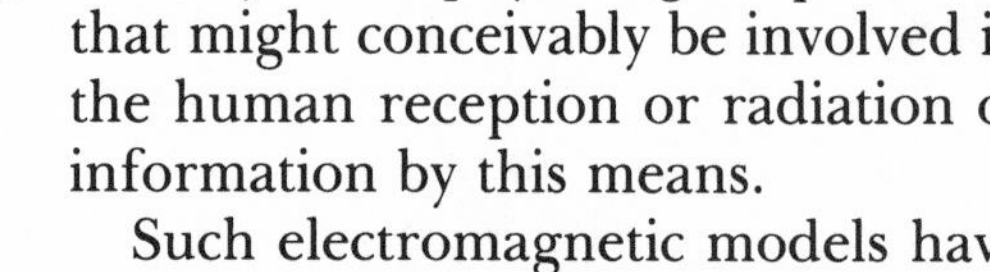

Such electromagnetic models have been sufficiently credible to stimulate numerous experimental tests, such as the use of elaborate "screen-rooms" or "Faraday cages" to enclose the operators or portions of the equipment, in attempts to shield or alter the strength or character of the electromagnetic signals purported to carry the anomalous information. Other studies have attempted to assess the decay in "signal strength" with distance; the "bit-rate" or "bandwidth" of information carried on the "channel"; or various other interference, diffraction, penetration, or polarization effects that would confirm or qualify the basic concept. Unfortunately, because of the very long wavelengths postulated and the uncertainties of the neurophysiological thresholds, such experiments have been unconvincing in establishing or rejecting this vehicle. In a related body of work, possible correlation of psychic effects with various forms of geomagnetic activity, including but not limited to electromagnetic phenomena, has been proposed and tested, again with unpersuasive results.

In our judgment, all such models founder on a more basic flaw: they have no capacity to accommodate precognition to the extent it appears throughout psychic research in general or in the specific precognitive remote perception studies described in Section III. Specifically, there is no mechanism in any of them that would permit information about a geographical target to be *advanced* in time by several days while it is delivered to a percipient thousands of miles away. Scrutiny of many other types of physically deterministic models inevitably leads to the same obstacle: each is incompetent to accommodate precognition as an element of the anomalies package.

Failing the capability of any straightforward transpositions of established physical theory, the only recourse is to move on, with great caution, to models that entail some basic expansion of the established scientific paradigm. Here, also, a variety of options may be contemplated, and a number of these have been developed to some extent. For example, attempts have been made to ascribe to consciousness an entropy-reducing capability that allows it to exert an ordering influence on otherwise random physical processes, thereby reversing their normal thermodynamic tendency toward minimum information and maximum chaos. All of the REG and RMC results described in Section II are in some sense consistent with such an ordering process; statistically speaking, the PK distributions carry slightly more information than the chance expectations or the calibration data. The PRP effects of Section III could also be regarded as a marginal ordering of the perception descriptor digits compared to their chance correlations with those of the targets. Although we have not attempted to develop this idea in analytical detail, the conceptual vocabulary it offers has proven useful in certain aspects of our work, and we shall return to it in Section V.

Another approach has been to recast basic physical laws in mathematical coordinate systems of more dimensions than the usual three space and one time components in which human experience has traditionally been represented. By carrying additional, intangible dimensions through such formalisms, various new terms arise in the results that can be assigned to represent anomalous effects without formal alteration of the physical laws themselves. For example, an imaginary component may be added to each of the real spatial and temporal coordinates to form a complex space and time metric, somewhat in the spirit of a.c. circuit theory, complex Fourier analysis, or exponential wave mechanics. These imaginary components and their interactions with the usual real components then permit accommodation of a family of anomalous effects not covered by the real terms alone.

Even more extreme coordinate transformations have been proposed based on the radical but intriguing hypothesis that universal information is most basically arrayed not in terms of position and time at all, but rather as frequencies and amplitudes of some abstract wave-mechanical system. In this view, human consciousness performs Fourier transforms on this frequency/amplitude information to convert it to the familiar space-time array on which our cognitive processes have come to rely. In the sense that physical space and time are thereby downgraded from fundamental coordinates of experience to merely useful ordering parameters, the possibility exists that consciousness, by assembling the appropriate frequencies and amplitudes, may construct information about *any* portion of space and time for interpretation in its conventional form.

Probably the most exercised category of radical model attempts to apply the concepts and formalisms of quantum mechanics to explain the anomalies encountered in modern psychic research. Quantum theory rests on an array of empirical postulates that contradict conventional logic, and it yields a corresponding array of predictions that are at variance with common impressions of reality. Among these are the basic quantization process that limits measurable properties to discrete values; the wave/particle duality; the role of the observer; the uncertainty, indistinguishability, complementarity, and exclusion principles; and the commitment to fully probabilistic mechanical behavior. All of these concede a degree of paradox in human perception of physical processes and suggest that physical theory is less a statement of abstract reality than of our ability to acquire information about that reality. Despite the enticing possibility of correlating such physical anomalies with consciousness phenomena, direct application of quantum mechanics to psychic processes has so far failed to yield completely credible, or even readily testable, hypotheses for the psychological, neurophysiological, and physical interface mechanisms. Rather, the formalism seems more profitably employed as a comprehensive metaphor for all forms of consciousness interaction, and this indeed will be a major tactic in our own approach.

All of these more severe attempts to extend physical conceptualization and formalism to encompass consciousness-related anomalies have merit in stretching the scope and arsenal of contemporary scientific theory beyond its established limits. However, they all suffer from a dearth of directly pertinent empirical data and an inability to incorporate certain subjective

or "soft" properties that must somehow be represented and computed if the models are to have pragmatic utility. In the following pages, we propose yet another form of radical model that draws insight from many of these versions, from more abstract philosophical perspectives, and, of course, from our experimental evidence, to propound a metaphor that both serves our specific anomalies research and opens options for broader comprehension of the nature and role of consciousness in general.

2

REALITY RECIPE

Our proposed model is built upon three basic postulates that might be termed the geometry of reality, the wave nature of consciousness, and the quantum mechanics of experience. The first of these is simply the conceptual configuration illustrated in Fig. IV-1, where consciousness is depicted by a diffuse sphere, surrounded by an otherwise boundless domain representing the environment in which that consciousness is immersed. Our use of the term *consciousness* is intended to subsume all categories of human experience, including perception, cognition, intuition, instinct, and emotion, at all levels, including those commonly termed "conscious," "subconscious," "superconscious," or "unconscious," without presumption of specific psychological or physiological mechanisms. The concept of *environment* includes all circumstances and influences affecting the consciousness that it perceives to be separate from itself, including, as appropriate, its own physical corpus and its physical habitat, as well as all intangible psychological, social, and historical influences that bear upon it. Thus, consciousness and environment engage in the "I/Not I" dialogue of classical philosophy, but with the interface between the two regarded as subjective and situation-specific.

It is a major hypothesis of the model that *reality*, encompassing all aspects of experience, expression, and behavior, is constituted only at the interface between consciousness and its

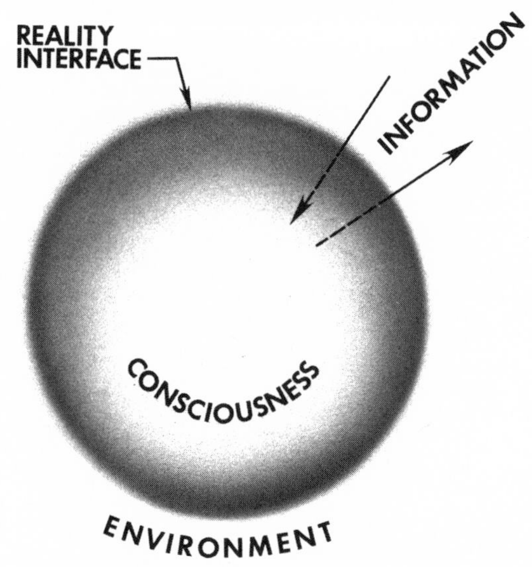

Fig. IV-1 *The Geometry of Reality*

environment, and it is to that reality interface alone that the model pertains. It is further presumed that the sole currency of any reality is *information*, which may flow in either direction; that is, consciousness may insert information into its environment as well as extract information from it. In this functional sense, information may be constituted by any array of stimuli that the consciousness, or the environment, is capable of sensing and reacting to.

Maintaining this perspective, we further propose that the purpose of *any* physical theory, or of any other scheme of conceptual organization, is to represent, organize, correlate, and predict the experiences of human consciousness interacting with its environment. Once again, neither that environment nor that consciousness can productively be represented in isolation; only in the interaction—in the exchange of information between the two—are palpable effects constituted. If the physical or physiological mechanisms that generate, convey, and accept information at the consciousness/environment interface are identifiable and predictable, the overall process is regarded as "normal." If any of these mechanisms are unknown, however, the process is "anomalous," and the prevailing theory must be regarded as inadequate or incomplete. This is the situation we now face with respect to the man/machine anomalies and the PRP information acquisition described in the previous sections.

It is also essential to our approach that the common concepts of established physical theories, such as mass, momentum, and energy; electric charge and magnetic field; frequency and wavelength; the quantum and the wave function; and even distance and time, be regarded as no more than useful information-organizing categories that consciousness has developed for ordering the chaos of stimuli bombarding it from its environment, or passing from it to its environment. As such, these concepts reflect as much the characteristics of consciousness as of the environment. More precisely, they reflect the characteristics of consciousness interacting with its environment. It follows, then, that the concepts, formalisms, and imagery of physical theory may provide useful metaphors for the representation of the nature and processes of consciousness itself or, again more precisely, of consciousness examining itself. Conversely, the less tangible aspects of subjective and impressionistic experience may be requisite ingredients of any general theory of reality, including physical reality.

"By using the processes of Nature as metaphor, to describe the forces by which it operates upon and within Man, we come as close to describing 'reality' as we can within the limits of our comprehension. . . . In this way, Man's imagination and intellect play vital roles in his survival and evolution."

JONAS SALK

The perspective implicit in these hypotheses is certainly not original; allusions to it may be found in the writings of many major scientists and philosophers of past and present eras.

Consciousness Contemplating Itself

During the emergence of science as a rigorous discipline, Francis Bacon suggested that

> all the perceptions both of the senses and the mind bear reference to man and not to the universe, and the human mind resembles those uneven mirrors which impart their own properties to different objects, . . .[2]

More recently, the astrophysicist James Jeans repeated the insight equally poetically:

> The concepts which now prove to be fundamental to our understanding of nature . . . seem to my mind to be structures of pure thought, . . . the universe begins to look more like a great thought than like a great machine.[3]

Similar ideas have been heard from other sectors; the founder of psychoanalytic theory, Sigmund Freud, defined "the test of science" to be

> fully circumscribed if we confine it to showing how the world must appear to us in consequence of the particular character of our organization. . . . [since] our mental apparatus . . . itself is a constituent part of that world which we are to investigate.[4]

And the philosopher Arthur Schopenhauer insisted that

> a consciousness without object is no consciousness at all. . . . Although materialism imagines that it postulates nothing more than this matter—atoms for instance—yet it unconsciously adds not only the subject, but also space, time, and causality, which depend on special determinations of the subject. . . . *the intellect and matter are correlatives,* in other words, the one exists only for the other; both stand and fall together; the one is only the other's reflex. They are in fact really one and the same thing, considered from two opposite points of view; . . .[5]

Corresponding persuasions of many of the patriarchs of modern physics were mentioned in Section I-6. To these might be added Einstein's classic reminder:

> Concepts which have been proved to be useful in ordering things easily acquire such an authority over us that we forget their human origin and accept them as invariable.[6]
>
> The system of concepts is a creation of man together with the rules of syntax, which constitute the structure of the conceptual systems. . . . All concepts, even those which are closest to experience, are from the point of view of logic freely chosen conventions, just as is the case with the concept of causality, . . .[7]

Regardless of their source, if one accepts the essence of such premises, it follows that any physical formalism may conceivably be adapted, via metaphor, to represent the characteristics of consciousness interacting with its environment. Actually, such metaphorical representation need not even be restricted to physical or indeed to general scientific vocabulary or concepts, but could derive its analogies from any organized intellectual schema. For reasons developed in the following two chapters, we shall employ the general formalisms of physical wave mechanics and the particular propositions of quantum wave mechanics.

"The same organizing forces that have shaped nature in all her forms are also responsible for the structure of our minds."

WERNER HEISENBERG

3

CONSCIOUSNESS PARTICLES AND WAVES

Historically, the distinction between the behavior of discrete, tangible blocks of matter—particles—and the diffuse, undulating behavior of various wave processes has permeated virtually all attempts of human consciousness to represent the mechanics of its physical environment. Certainly early man encountered one portion of his pragmatic reality in particulate terms. His hand, his axe, and his cave; the trees, animals, and birds; the sun, moon, and stars were all sharply defined objects among which specific relationships could be visualized and implemented. Yet he doubtless noticed and pondered various wave-like processes as well—on the surface of a brook or ocean, in the cirrus patterns of the sky, or in the wind-driven vibrations of the leaves—and frequently represented these in his art and rituals, perhaps better to comprehend their mechanics as they bore on his maritime or agricultural activities, or perhaps to invoke some of their aesthetic and mystical characteristics.

From these earliest impressions, the concepts of particles and waves have evolved to considerable philosophical and analytical sophistication as two modes of representation of a broad variety of physical phenomena. In numerous cases, one of these perspectives has remained dominant for some period, subsequently to be replaced by the other when proven inade-

Anasazi Petroglyphs (Eastern Arizona)

quate to accommodate some new body of empirical evidence, and then again to be joined by the former in some dualistic compromise. Early Greek scholars were driven to propose a wave theory of sound to explain certain observations that were inconsistent with the prevailing atomistic perspectives. Many centuries later, a growing body of anomalous experimental data forced development of a wave theory of light to challenge, and eventually to displace, Newton's well-entrenched "corpuscular" model. Toward the close of the nineteenth century, the derivation from Maxwell's field equations of testable quantitative properties of electromagnetic waves subsumed particulate electrodynamics and unified much of prevailing physical theory in what then appeared to be a comprehensive philosophical fashion.

But particulate light, sound, and electricity were by no means dead and, conversely, particulate substance was soon to be challenged. Twentieth-century physics presented a plethora of new paradoxes, including the photon, the phonon, and the quantum, and, most perplexing of all, the wave nature of matter. Blackbody radiation and the photoelectric effect could not be explained by the wave theories of light; the Planck-Einstein frequency-quantized, particulate photon was essential to their explication. On the other hand, rationalization of the Bohr-Sommerfeld atomic energy levels and of the experimental data on electron diffraction rested on the monumental proposition from de Broglie and Schrödinger that particulate material objects could, in some situations, display wave-like characteristics.

To this day, these wave/particle paradoxes have not been fully resolved in any convincing philosophical sense. For practical purposes, science uncomfortably concedes that under certain circumstances, light and matter behave like discrete particles; under other circumstances, they behave like waves. Analytically, we can derive useful and accurate predictions of their behavior,

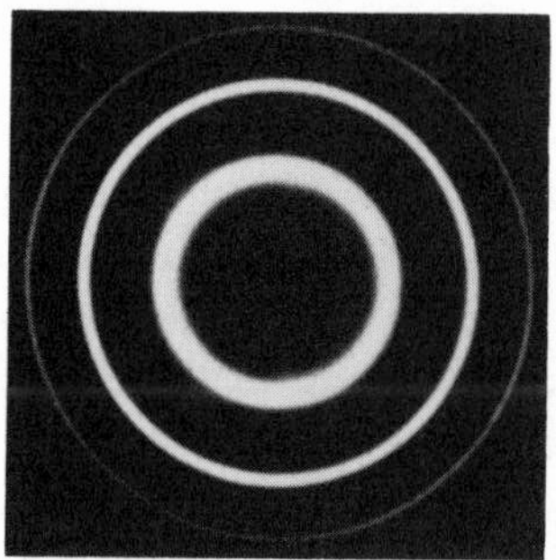

Optical Interference Pattern

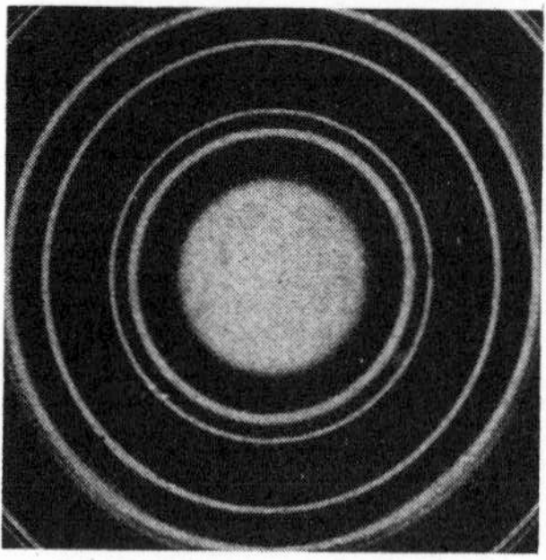

Electron Diffraction Pattern

if we can first guess which ethic will prevail in a given situation. To some extent we can even force some coherence of these perspectives by composing particle-like "wave packets" or waves of particulate probabilities, but philosophically the two formats must be regarded, in Bohr's profound terminology, as "complementary." In other words, the wave and particle perspectives are not mutually exclusive; rather, some mixture of both is necessary to represent the phenomena fully.

Our interpretation of this irreducible complementarity, consistent with the postulates of the model, is that it is not the physical world *per se* that imposes such dichotomy; it is our consciousness. More precisely, it is imposed by the process of consciousness interacting with its physical environment. And if this interpretation is valid, then it is also consistent to attribute to consciousness itself the option of wave-like as well as particulate behavior, again as perceived by that same consciousness. In other words, the consciousness that has conceived both particles and waves, and found it necessary to alternate them in some complementary fashion for the representation of many physical phenomena, may find a similar complementarity necessary and useful in representing itself.

"It seems at least conceivable that what is true of perceived objects may also be true of perceiving minds; just as there are wave-pictures for light and electricity, so there may be a corresponding picture for consciousness."

JAMES JEANS

Pursuing this line, we would then note that over most of the history of Western thought, and no less so today, the commonly prevailing conceptualization of consciousness has been basically "particulate" in nature. That is, an individual consciousness has usually been presumed to be well localized in physical space and time, namely in its host physiological corpus, and to interact only with a few aspects of its environment and a few other similarly localized consciousnesses at any given point in its experience. Thus, in this view, any society of consciousnesses could be said to resemble an ensemble of gas molecules in some elaborate environmental container, bouncing around against their immediate neighbors and against the proximate environmental features in some sort of grand gas-kinetic dance.

So long as this particulate paradigm is retained, effects like the remote influence on physical behavior or the remote acquisition of information observed in our experiments must remain inherently anomalous. But if consciousness were to allow itself the same wave/particle duality that it has already conceded to numerous physical processes, these situations would become more tractable. In particular, a wave-like consciousness could employ a host of interference, diffraction, penetration, and remote influence effects to achieve normally most of the anomalies of the particulate paradigm.

To develop such a wave mechanics of consciousness, it will be necessary to specify more precisely the salient properties of such waves, and this task may be facilitated by brief review of the nature and treatment of more familiar wave phenomena. Natural wave processes are usually discussed in terms of their physical class, the medium in which they propagate, the particular property of the medium that oscillates, the factors driving and restraining their oscillations, the geometry of the oscillations, the speed of propagation of their wave fronts, their rate of attenuation, and numerous other physical and geometrical factors appropriate to the situation in which they are expressed. For example, the vibration of a taut string entails periodic displacements perpendicular to its alignment, initiated by some external agent, such as a plucking finger or a mechanical shaker, and restrained in amplitude by the tension in the string and by its mass. Sound waves consist of alternating patterns of compression and expansion of the ambient gas, liquid, or solid in which they are propagating, again initiated by some external vibrator and restrained by the compressibility of the medium. Water waves are marked by regular or irregular undulations in the height of the free surface, driven by wind or by subsurface disturbances and restrained by gravitational or surface tension effects. Electromagnetic waves, of which light

Propagating Water Waves

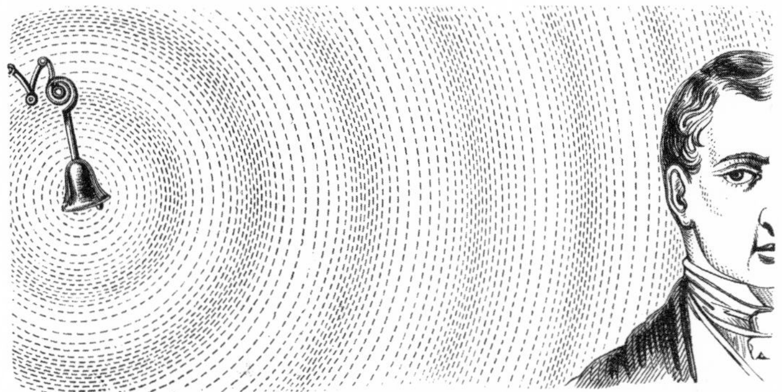

Propagating Sound Waves

is a special category, involve oscillations of the local electric and magnetic field magnitudes propagating in a tightly coupled interaction as constrained by certain constitutive properties of the medium. Depending on the regularity of the source and the geometry and uniformity of the medium, any of these waves may be simple sinusoidal trains of uniform amplitude, wavelength, and speed of propagation, or they may be more complex in their wave shape and propagation characteristics.

When any wave train reaches a boundary or discontinuity in the medium in which it is propagating, some portion of it is reflected and some portion is transmitted. If all relevant boundaries of the medium are fixed in some fashion, the succession of incident, reflected, and re-reflected propagating waves compound to establish a pattern of "standing" waves, wherein the displacements of the medium wax and wane in a stationary, rather than a propagating configuration. For example, a violin string pinned at both ends and forcibly oscillated by a bow at some intermediate point establishes a configuration of standing waves having regularly spaced nodes and antinodes (positions of zero and maximum displacement). Since nodes must be maintained at the fixed ends, only a discrete set of patterns can be sustained by a given length of string. The simplest of these patterns, called the fundamental, has just one antinode, midway along the string. The others, called overtones or harmonics, have, in progression, two, three, four, etc. equally spaced antinodes. The vibration frequency of any pattern depends on the distance between nodes, the mass-per-unit length of the string, and the tension under which it is held. The quality of the tone emitted by the violin depends on the particular mixture of overtones superimposed on its fundamental vibration.

In similar fashion, a reverberating tympani head, or the surface of water in a vibrating teacup, displays circular or zonal standing wave patterns, depending on the mode of agitation. Even more complex two- and three-dimensional patterns can

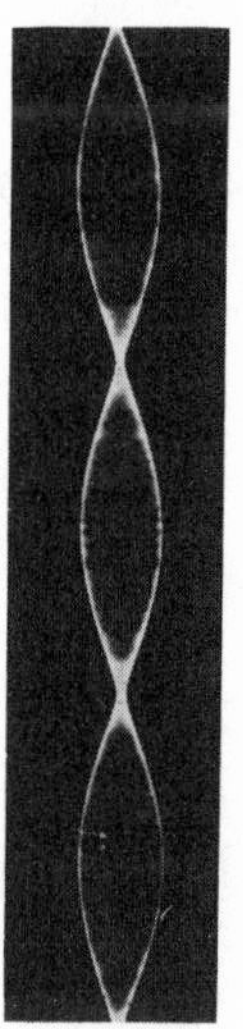

Standing Waves on a String

Chladni's Figures of Standing Wave Patterns on a Plate

arise in acoustical cavities, such as woodwind or brass musical instruments, or in various electromagnetic resonators, such as wave guides, transmission lines, and optical interferometers. In all cases, these characteristic standing wave patterns are products of four interacting influences: the class of wave phenomenon, the nature of the medium supporting the wave, the boundary conditions or constraints imposed on the medium, and the driving source. If any of these is altered, the standing wave pattern reflects that alteration.

In either the propagating or standing form, wave processes enjoy a number of capabilities that are not available to discrete particles. Most basic is that the information and energy they carry may be widely diffused over broad regions of space and time, rather than sharply localized in a well-defined geometrical region. For example, a simple point antenna will indiscriminately collect information from, or radiate information to, all of its spherically surrounding space. Only by more elaborate configurations, such as a radar dish or a conical megaphone, may the focus of its collection or radiation be narrowed to better defined remote regions.

Unlike particulate phenomena, where colliding objects distinctly alter one another's trajectory, intersecting waves may pass through each other with no permanent distortion, yet during that intersection complex superposition or "interference" patterns may be formed. While particulate beams cast shadows that are sharply defined by intervening obstacles, waves have the capacity to diffract around corners and thus access regions precluded from rectilinear motion. The diffraction of sea waves around a jetty or through a harbor gateway are familiar examples, as are the echoes of sound along confined passages, or pinhole optical effects. The ability of waves reaching an interface or discontinuity in the medium to reflect some portion of their amplitude and transmit another portion also differs sharply from particulate collisions wherein the incident particle, like a marble in a soup bowl, either surmounts the interface intact, or is totally reflected. Of special use in our metaphor are so-called "evanescent" waves whose influence can penetrate for some distance into regions where they cannot propagate in their usual oscillatory modes, and from which their particulate counterparts would be completely excluded.

Note that for the simpler classes of physical waves, such as vibrating strings and membranes, sound, and even less regular forms like water and seismic waves, the specification of the wave properties and their relationship to the characteristics of the medium and source are conceptually clear. In other cases, however, some of these identifications and relations are more

obscure, and abstract argument must be accepted if the utility of the conceptualization is to be retained. For example, the evident ability of electromagnetic waves to propagate through a vacuum raises difficult questions about the properties of a total void that can support the oscillating stresses of the alternating electric and magnetic fields. Search for characteristics of this "ether" has occupied physicists for over a century, and the issue remains incompletely resolved.

Such empiricism becomes even more strained in the implementation of a wave mechanics of matter. De Broglie's original hypothesis was that physical substance could also display wave-like characteristics of interference, diffraction, and interface penetration. The wavelength implicit in such effects was postulated to be inversely proportional to the momentum of the associated particle, so that the greater its mass or velocity, the shorter its effective wavelength. Schrödinger was subsequently able to compute with impressive precision the energy levels of simple atomic structures under the hypothesis that the electrons were constrained to standing wave patterns consistent with the spherical boundary conditions imposed on them by the electric potential profile of the atomic nucleus. But both of these immensely productive propositions raised profound questions regarding the particular property of the physical object that was oscillating, as well as the properties of the medium supporting the oscillation. Despite the architecture of successful theory that has since been developed on the basis of such quantum wave mechanics, these fundamental questions remain philosophically troublesome and continue to be debated today.

One school of response—the so-called "Copenhagen" or "observational" position advocated by Bohr and his colleagues—takes the stance that the amplitude of these matter waves, when appropriately traced through the mathematical formalism, is indicative of the likelihood of observing the associated particle in a given state, at a given position and time, when an appropriate experiment to measure those properties is actually implemented. The process whereby the particle reaches that state and position, in the absence of any direct observation of such process, cannot properly be addressed via quantum theory, and indeed paradoxical inferences can arise if such interpretations are attempted. Hence, this wave mechanics of matter does not describe physical behavior-in-itself; it only describes the *observation* of physical behavior.

Heisenberg's interpretation of this fundamental change in perspective is consistent with the central premise of our proposed topology of reality:

> . . . in the Copenhagen interpretation of quantum theory we can indeed proceed without mentioning ourselves as individuals, but we cannot disregard the fact that natural science is formed by man. Natural science does not simply describe and explain nature; it is a part of the interplay between nature and ourselves; it describes nature as exposed to our method of questioning. This was a possibility Descartes could not have thought, but it makes the sharp separation between the world and the I impossible.[8]

So heuristic is this observational presumption that the quantum theory becomes essentially metaphorical in character; that is, the conceptual imagery of palpable wave processes is here being transposed to a far more abstract level—the observation of microscopic natural phenomena. Thus, like the electromagnetic and acoustical wave theories before them, the wave mechanics of matter is an attempt to represent effects we cannot directly observe by analogy to effects we can readily see, such as water waves or standing mechanical vibrations.

> **"The *possibility of experience* is, then, that which gives objective reality to all our *a priori* cognitions."**
> IMMANUEL KANT

Even among aficionados of the Copenhagen school, there is a major division on the implications of this posture for the role of human consciousness. Some vigorously contend that observation can be achieved without involving consciousness as an essential element of the system; others—and we shall stand in this camp—feel that consciousness, defined in proper generality, is the ultimate factor in any observation, and therefore an intrinsic ingredient in the conceptualization and formulation of matter waves.

In either interpretation, the Copenhagen precedent nicely serves the central postulate of our consciousness model. Namely, if consciousness finds wave mechanics a useful complement to particulate physics in representing a number of physical processes like sound, light, and the atomic-scale behavior of matter, and if it is willing to tolerate increasingly more abstract and metaphorical specification of the wave properties as it progresses to subtler phenomena, then consciousness may also find a wave-mechanical metaphor to be conceptually and functionally useful for representing itself. The precedent is also particularly helpful in suggesting the nature and substance of these consciousness waves. Namely, if we generalize the concept of "observation" to encompass all information-processing capacities of consciousness as well—cognition, emotion, behavior, and any other form of experience—then the same pragmatic interpretation may obtain. So, just as the Copenhagenists speak of "probability-of-observation" waves, we shall postulate "probability-of-experience" waves.

4

QUANTUM CONCEPTS

Beyond the various departures from classical particle behavior implicit in its wave-mechanical approach, the science of quantum physics imposes a number of other empirical postulates that appear to contradict traditional human experience, yet are necessary to accommodate experimental data. It is our position that these postulates are also indicative of basic human information acquisition and processing mechanisms and thus can provide further useful metaphors for representing them. Chief among these is the identification of a fundamental unit of quantization, called Planck's constant and usually denoted by the symbol h, that appears in several crucial capacities. For example, it relates the wavelength of de Broglie's matter waves, λ, to the associated particle momentum, p, in the form

$$\lambda = h/p \; . \tag{4.1}$$

It also defines the minimum indivisible unit or quantum of energy, E, associated with electromagnetic oscillations of frequency ν:

$$E = h\nu \; . \tag{4.2}$$

In yet another form, termed the "quantum of action," Planck's constant establishes an ultimate granularity to all physical observations, and thus, in our metaphor, to all experiences of consciousness in general. Because of its extraordinarily small size—6.626×10^{-34} joule-sec in the most common physical metric—this quantization is not perceptible in the great bulk of macroscopic physical experience, but its numerous pragmatic and philosophical ramifications on the atomic scale may nonetheless provide valuable analogies for the conceptualization and representation of some of the subtler processes of human consciousness. Quantitative evaluation of this unit of experience in these more general exercises of consciousness will depend on the particular metric schemes employed.

Also devolving from this intrinsic granularity of observation are a number of quantum mechanical principles that further delimit the applicability of classical mechanics to atomic-scale situations and can also prove metaphorically useful in the consciousness application. One of these derives from the classical concept of "conjugate coordinates"—orthogonal pairs of physical properties that serve to define the dynamical state of a given object or system. The most common conjugate pairs are the components of linear momentum and their corresponding position coordinates; angular momentum and the corresponding angular position; and energy and time. In the classical particulate realm, it is presumed that such pairs of properties can be simultaneously specified and measured with arbitrarily high precision. If one is committed to a wave-mechanical representation of physical matter, however, inherent limitations on the simultaneous specification of such conjugate properties must be conceded, in a form first posed by Heisenberg and called the *Uncertainty Principle*:

$$\Delta p \Delta q \approx h \,. \qquad (4.3)$$

Here Δp and Δq denote the uncertainty in determination of each of the properties, and h again is Planck's constant. After pondering the philosophical implications of this constraint, Bohr broadened the concept to a *Complementarity Principle* which recognized that not only the mechanical measurables, but all conceptual perspectives employed to localize the state of a system are subject to comparable limitations on mutual precision. In this broader form, the principle is directly applicable to our purposes as an empirical constraint on the ability of consciousness to perceive and represent any complementary aspects of its experience.

The *Correspondence Principle*, first posed by Planck and later clarified by Bohr, acknowledges the necessity for the quantized behavior of atomic-scale systems to blur into the geometrical and dynamical continua of common experience as the dimensions approach the macroscopic. In essence, this statement argues that whenever the de Broglie wavelengths of the systems under study are very small compared to the scale of observation, classical behavior should be a satisfactory approximation to the totality of the very fine-grained quantum effects. Thus, quantum mechanics is regarded as the more fundamental formalism, subsuming classical mechanics as a special case that applies whenever very large numbers of quanta are involved. For our purposes, the interpretation will be that while consciousness acquires and processes information about the physical world in discrete units, in most situations these units are so small and abundant that there is an illusion of a continuum, as in a stream of sand or a motion picture.

"If the aim of analysis is to separate, it must stop short of the ultimate structural units; because when the parts become so simple that they are indistinguishable, their indistinguishability confuses them in our observational knowledge and, in a sense, undoes the separation which the analysis has effected."

ARTHUR EDDINGTON

Consistent with the Copenhagen interpretation of quantum mechanics is the *Indistinguishability Principle*, which affirms that no physical observation can distinguish between two identical interacting atomic units. For example, theoretical representation of the hydrogen molecule must acknowledge that no measurement can distinguish which of its two electrons is most directly coupled to one or the other of its two nuclei; it can only establish that one of the two electrons is in state *a*, the other in state *b*. This indistinguishability logic, when carried through the formalism, has major practical consequences for the strength of the molecular structure; in fact, it is the source of a classically anomalous binding effect called the "exchange energy" that is the dominant contribution to the covalent chemical bond. The associated philosophical point, important to our argument, is that the surrender of *information* distinguishing the identities of two interacting subsystems translates into an increment in the structural *energy* of the bonded system.

Related to the indistinguishability concept, yet more extensive in its implications, is the *Exclusion Principle* posed by Pauli, which precludes any two interacting electrons from existing in precisely the same state. This forces postulation of an electron "spin," or internal angular momentum, that is itself a metaphorical transposition of a classical concept. From broad experimental evidence it appears that this electron spin can take only two possible values: $\pm$ ½ unit of $h/2\pi$. In other words, the electron acts as if it were spinning about some given axis at some fixed rate in either a clockwise or counterclockwise sense. The exclusion principle then permits two electrons to occupy otherwise identical states if their spins are of opposite sense, but not if their spins are in the same direction. This, in turn, predicates population of the energy levels of atomic and molecular structures with pairs of oppositely spinning electrons, a propensity that is reflected in the physical and chemical properties of those elements and compounds. The exclusion principle also has major implications for the statistical behavior of large numbers of interacting identical particles such as the conducting electrons in a metal. Association of this already highly metaphorical spin property with the bonding tendencies of human consciousness is an important feature of our model.

With these various principles overlaid on the formalisms of classical wave theory, quantum wave mechanics proceeds to its task of computing observable atomic-scale properties by incorporating the postulates of de Broglie and Einstein, Eqs. (4.1) and (4.2), into a generic wave equation, first proposed by Schrödinger:

$$\left[\frac{h^2}{8\pi^2 m}\right] \nabla^2\psi = (V - E)\psi \qquad (4.4)$$

where h is again Planck's constant, m is the mass of the object represented, ψ is the probability-of-observation amplitude function, V is the potential energy environment, E is the total energy of the object, and ∇^2 denotes the second spatial derivative or Laplacian operator. This equation effectively replaces the classical particulate dynamics of the object by patterns of probability-of-observation standing waves that are consistent with the imposed environmental constraints. For most atomic systems, these constraints are provided by the electrostatic forces between the nuclei and electrons and may be represented in the form of three-dimensional "potential wells" that are somewhat akin to the teacups or cavities that confine standing water, acoustic, or electromagnetic waves. For example, in a hydrogen atom the single electron is constrained by the electrostatic field of the single nuclear proton. Even for this simplest of atomic configurations, the permitted standing wave patterns are quite complex, consisting of a hierarchy of so-called spherical harmonics compounded with radial variations known as "Laguerre polynomials." These complex nodal patterns—called *eigenfunctions*—are indexed by three sets of running integers, or "quantum numbers," that are empirically associated with the energy of the particular state, its total angular (rotational) momentum, and its angular momentum about the polar axis of the configuration (or, equivalently, its magnetic moment), respectively. Thus, these quantum numbers relate the mathematical representation of the atomic structure to its observable physical properties. Overlaid on all of this is the binary option of the electron spin.

"The ψ function, in fact, does not represent something which would have its place in a point of space at a given instant; it represents, taken in its entirety, the state of knowledge of an observer, at the instant considered, of the physical reality that he studies; there is nothing surprising, therefore, in the fact that the function varies from one observer to another."

LOUIS DE BROGLIE

Computation of the observable states of more complex atomic

and molecular structures can be approached in similar fashion by first specifying the potential well of the appropriate nuclear charge configuration, inserting the requisite number of electrons, acknowledging their own electrostatic interactions, spin alignments, and exchange effects in some iterative schema, and again searching for self-consistent patterns of standing waves. These patterns can become extraordinarily complex, but the identification of the quantum numbers that relate them to observable physical properties remains central to the method.

The transition of an atomic or molecular structure from one state to another may be accomplished by direct collision with neighboring atoms or molecules, or by the emission or absorption of a quantum of electromagnetic radiation carrying precisely the requisite amount of energy differential, in accordance with relation (4.2). Such interactions constitute important mechanisms of communication with, and influence on, other atomic units, and thus help determine the macroscopic characteristics of any large collection of such units.

Extension of quantum mechanical formalism to ensembles of more than a few interacting atomic systems becomes progressively more cumbersome analytically; for large numbers, statistical methods are required. As noted earlier, all statistical predictions ultimately trace back to the basic probability rules imposed on the elemental units. For example, classical statistics rests on a particular set of presumptions regarding *a priori* probabilities of specific dynamical events. In the quantum domain, however, other constraints, most especially the exclusion principle, complicate the elemental probabilities and can lead to radically different statistical results. Thus, classical statistics stands as a special case of quantum statistics that obtains under most macroscopic conditions. For our purposes, this raises one more metaphorical precedent, namely, that human experience, as established by the interaction of consciousness with its environment, may also play by more elaborate probability rules than commonly presumed and hence may require a more complex statistical treatment to represent it appropriately.

Many other aspects of quantum mechanics could be proposed for metaphorical representation of the experiences of consciousness, but our purposes will be better served by developing the particular analogies already suggested to some conceptual and practical utility. One crucial preparatory step remains, however: the identification of a consciousness metric.

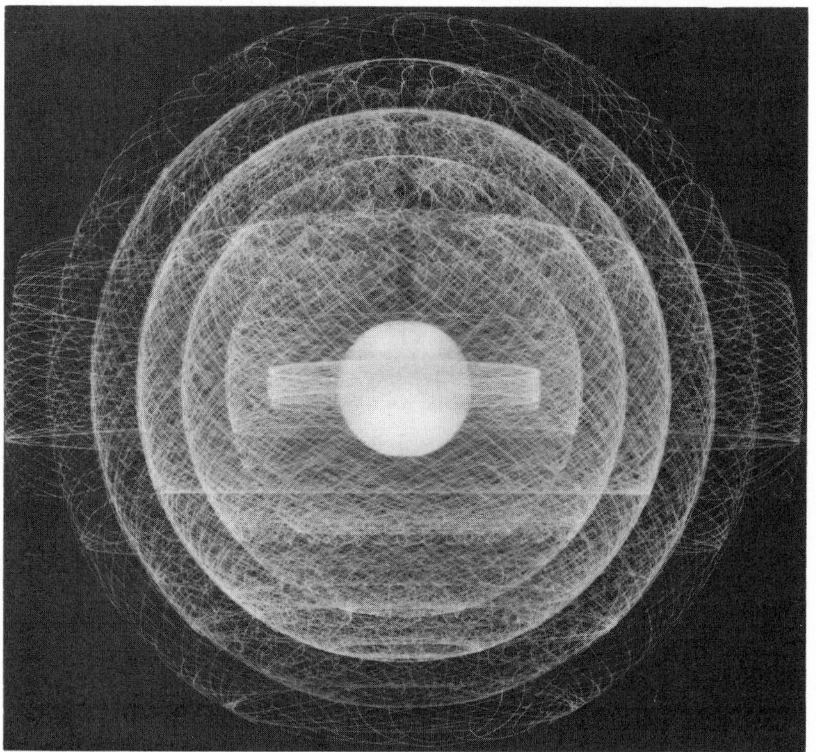

Time-exposed Photograph of an Animated Model of a Uranium-235 Atom

5

CONSCIOUSNESS COORDINATES

Before attempting to transpose quantum wave mechanics to a more general representation of the affairs of consciousness, some system of coordinates and metric properties must be established whereby such processes can be quantified. In the conceptualization and formulation of the mechanics of the physical world, certain basic metric properties like distance, time, mass, and electric charge have traditionally been used along with a variety of quantities derivable from them, such as velocity, momentum, energy, force, magnetic moment, and frequency. These particular choices have evolved diffusely over the history of science, fostered by combinations of pragmatic measurability and conceptual appeal that allowed effective consensus application. The selection of a more comprehensive set of coordinates and properties in which the full range of consciousness experience may be represented is complicated by the broader variety of effects that must be accommodated, and especially by their more subjective and aesthetic features.

> **"Quantities like length, duration, mass, force, etc. have no absolute significance; their values will depend on the mesh-system to which they are referred. . . . There is no fundamental mesh-system. . . . The systems used in current physics are arbitrary."**
>
> ARTHUR EDDINGTON

To begin this task, we again note that the concepts utilized in physical theory are themselves creations of consciousness and hence may reflect its fundamental propensities when ordering its acquired information in any area. In other words, our tried and trusted indicators of physical experience may merely be special cases of some grand organizational metric inherent in the nature of human consciousness.

Some support for this hypothesis and some hints for its development may be found in the principal mode of expression of consciousness, its language. The primary value of any concept is the degree to which it facilitates comprehension of a large number of situations for a broad range of individuals. Thus, a search of the linguistic lexicon for basic conceptual associations and metaphors that consciousness finds serviceable in communicating the essence of its experience may reveal aspects of its primal metric structure. Of particular interest for this purpose are the physical metaphors it verbalizes to describe its own states.

"All our ordinary verbal expressions bear the stamp of our customary forms of perception."

NIELS BOHR

Some contemplation of these historical and linguistic precedents suggests the following consciousness metric:

Consciousness Distance

In the physical domain, specification or measurement of distance, length, or location in essence constitutes response to an explicit or implicit inquiry of "where?" or "how far?" If this response is merely impressionistic, for instance, "nearby," "far off," or "over there," little analytical logic can follow. To quantify the situation requires an agreed-upon standard of length. Historically, such standards have progressed from very crude indicators, such as the length of a typical human foot or stride, to progressively more refined rulers, to precisely marked and carefully maintained precious metal bars, to the wavelengths of particularly stable atomic spectral lines. But at every stage the quantification of physical distance has been a statement of

Quantification of Distance (16th Century)

relation to a given object or process, rather than an absolute matter.

To extend this concept to more general activities of consciousness, we might note that impressionistic verbal allusions to physical distance are commonly used to describe all manner of cognitive or emotional situations. A person is described as a "close" friend or "distant" relative, as "deep" in thought or "high" as a kite; an idea may be "central," "remote," or "far out"; and we allow our minds to "wander" over various conceptual "grounds," before taking a "position" on an issue. Lucid as these analogies may be, they are inherently qualitative; to quantify such consciousness dimensions, some referential standard must be established—some conceptually evident "stride," "meter," or "wavelength," as it were—in terms of which consciousness "distance" may be specified in quantitative form.

At this primitive point in our understanding of consciousness mechanics, no such precise standard is evident, and we have little recourse but to revert to the same metaphor used for the earliest physical measurements, namely the number of "steps" or "strides" of consciousness required to progress from a given origin to a given goal. These may be cognitive steps of deductive logic, or complementary steps of emotion, intuition, or sensation, but all entail incremental acquisition of information relevant to the purpose. Collectively, these steps form a grid of experience, along which consciousness proceeds toward its goal by making discriminations or associations at each juncture, much as in a puzzle maze or in the games of "Twenty Questions" or "Charades." Consciousness distance, then, is quantified by the number of such steps required to transfer from one experiential position to another.

The concept of quantized subjective experience is not new. William James proposed a similar mechanism many years ago in terms of

> finite and perceptible units of approach—drops, buds, steps, or whatever we please to term them, of change, coming wholly when they do, or coming not at all. Such seems to be the nature of concrete experience, which changes always by sensible amounts, or stays unchanged.[9]
>
> The simplest thing, therefore, if we are to assume the existence of a stream of consciousness at all, would be to suppose that things that are known together are known in single pulses of that stream.[10]

Whatever their particular nature, these scaling increments of experience are inherently more subjective than their physical counterparts, and they may not share the latter's convenient mathematical linearity; in other words, simple associative, distributive, and commutative algebra may not apply. Possibly they may have a logarithmic or exponential character, as in some forms of information theory or in certain psychological models of perception or learning, or they may entail even more complex relationships, like the space-time distortions of general relativity. It also seems reasonable to suspect that this consciousness space grid, like the earliest physical length standards, varies somewhat from individual to individual, so that the number of bits of information required to accomplish a given experiential transition will also differ.

Consciousness Time

The identification and scaling of an analogy to physical time in the consciousness domain is equally important to the development of the metaphor, and even more elusive. Initial temptation simply to transpose this property directly—that is, to presume that the same clocks are relevant to all experiences of consciousness—should probably be set aside in the face of obvious temporal distortions that occur in both emotional and cognitive personal experience. In periods of satisfying creative work or pleasurable recreation, "time flies"; when the activity is tedious or boring, "time drags"; and in highly emotional situations, "time stands still." Dreams, mystical trances, and

The Cycle of Time (Ancient Mexican Calendar)

near-death experiences are frequently characterized by bizarre distortions or even total irrelevance of conventional time. Other allusions to time and temporal sequencing abound in common conversation: "time after time"; "time out"; "in the meantime"; "for the time being." In fact, not just the vocabulary, but even the conjugation patterns and syntax of most modern languages are so temporally presumptive that any attempt to communicate without explicit or implicit time references becomes chaotic. There are, however, some languages that are much less temporally structured, reflecting cultures that seem less committed to linear logic and sequential causality.

Questions of internal or consciousness "clocks," and of subjective versus objective qualities of time, have long attracted the attention of philosophers, biologists, psychologists, and physicists. While their far-ranging studies have explored many fascinating aspects of the topic, here we are concerned primarily with the need for some generalization of the concept of physical time to the consciousness frame, by which the temporal quality of experience may be metered in a more systematic, albeit intuitive, fashion. Physical time has traditionally been scaled by various regular natural phenomena, such as solar, lunar, and terrestrial cycles; burning candles, water drippings, sand flows, and pendulum oscillations; and most recently by precise molecular vibrations or atomic transitions. In the same spirit, we now search for a consciousness oscillation or processing cycle by whose period all cognitive and emotional experience may consistently be timed.

In the absence of more basic understanding, let us adopt as an empirical standard some posited consciousness scanning or pulsing frequency. Specifically, let us presume that consciousness, somewhat like a radar or sonar system, acquires experience of its environment by querying it, receiving response from it, and processing that response, all on some base frequency of operation. As with the distance scale, this scanning frequency probably varies from one individual to another, and possibly within the individual as well. In musical analogy, any given consciousness may process its experience in march time, waltz time, ragtime, or even hard rock, but nonetheless the associated periods of oscillation can serve to scale this dimension of experience in a self-consistent fashion. Thus, to express the rate of information assimilation in the consciousness domain, we form derivatives not with respect to physical time, but rather with respect to the prevailing consciousness time; that is, we specify units of information per processing scan of that consciousness.

A similar concept appears in a paper entitled "Oscillations as Possible Basis for Time Perception," by E. Pöppel:

> The results presented in this paper suggest that the temporal continuum is subjectively quantized into discrete units, which successively follow each other. . . . The quantization can formally be described as an oscillation.
>
> The duration of "quantum-period" seems to be influenced by several physiological conditions. . . . Individual differences can also be considerable. . . . The quantum-period discussed here possibly has an intimate relation to this temporal organization of language.[11]

Despite the vagueness of this concept of consciousness pulsation, with it so postulated, we can now invoke the Einstein-Planck relation, Eq. (4.2), to define a quantum of consciousness "energy," in terms of this frequency of processing of information. Thus, consciousness interactions of greater energy involve higher information-sampling frequencies, perhaps in somewhat the same fashion as food processors, lawn mowers, or voracious rabbits devour their respective fodder.

Consciousness Mass

In physical theory, the concept of *mass* functions in several contexts, in any of which it may be fundamentally defined and calibrated. For example, "inertial mass," as embodied in Newton's second law or in the Schrödinger equation, relates to the magnitude of the force necessary to accelerate or decelerate an object at a given rate. "Gravitational mass" is implicit in the force two objects exert upon one another at a distance, as represented by Newton's inverse-square law of gravitation or, in more sophisticated terms, by the distortion of the space-time grid prescribed by general relativity theory. In the high-energy domains of nuclear forces and transformations, particle "rest mass" and energy are interchangeable in accordance with Einstein's famous $E = mc^2$ equation. Although there are some fine theoretical points concerning the relationships among these various ramifications, for functional purposes any form of mass may be scaled by reference to a conceptual or empirical standard, such as one cubic centimeter of pure water, or a platinum-iridium cylinder maintained at the International Bureau of Weights and Measures.

Rhetorical allusions to similar concepts in the consciousness domain are familiar. We refer to a thought or experience as "heavy," to a person as "burdened," or to an opportunity or task as "massive," by which we imply that the importance, profundity, consequences, or "gravity" of the matter is sub-

Cosmic Mass: Siberian Chukchi Celestial Drawing

stantial. Conversely, we can treat an issue with "levity," or deport ourselves in a "lighthearted" manner. Thus we are acknowledging some property of experience that relates to the difficulty of maneuvering in consciousness space, or to the impact of a "matter" on the consciousness. We recognize that "light" thoughts or experiences readily drift away from the center of our individual consciousness and permit us to shift perspective more readily. "Heavier" matters are more difficult to advance or expel, tending to return inexorably to our cognizance where they "weigh" upon us. With allusion to the more elaborate concepts of general relativity, the most "massive" experiences are capable of distorting our consciousness perception grid and contextual framework. In the nuclear branch of the metaphor, a sudden reduction of consciousness mass may release copious energy for other purposes.

The quantification and standardization of the degree of consciousness mass in terms of some evident unit process proves even more elusive than for consciousness distance and time. Nonetheless, presumption of some such index, related to the amount of information or energy required to displace the consciousness from a given position, is essential to metaphorical generalization of other physical concepts and relationships.

Consciousness Charge

All physical processes subsumed within the topics of electricity and magnetism can ultimately be traced to the identification in nature of two classes of electric charge, negative and positive, as embodied by the electron and proton (or positron), respectively. Coulomb's law states that any two like charges repel, and unlike charges attract, with a colinear force proportional to the product of their charge magnitudes and inversely proportional to the square of their separation—a functional relationship very similar to Newton's law of gravitation. Standardization of the unit of charge may be variously established by reference to Coulomb's law, by an electrochemical recipe, or by the fundamental electronic particle itself.

Charge may be placed on or in a physical object by various processes, such as direct transfer or induction from another charged body, triboelectric rubbing with some complementary material, bombardment by a particle beam, or electrochemical reaction. Once configured, given geometrical arrangements of charge establish an electrostatic force field among them and store electrostatic energy. If the charges are free to move in response to these forces, their motion constitutes electric currents, which in turn are the sources of magnetic fields that can exert another pattern of forces on the currents. From the ensemble of charges, currents, electric and magnetic fields, and an assortment of relevant properties of the materials in which they are immersed, all of the phenomena of classical electromagnetism then follow.

Generalization of these concepts to the consciousness domain is again suggested by vernacular allusions to "charged" situations, feelings of "attraction" or "repulsion," "currents" of thought, "polarized" issues, and so on. For example, we rec-

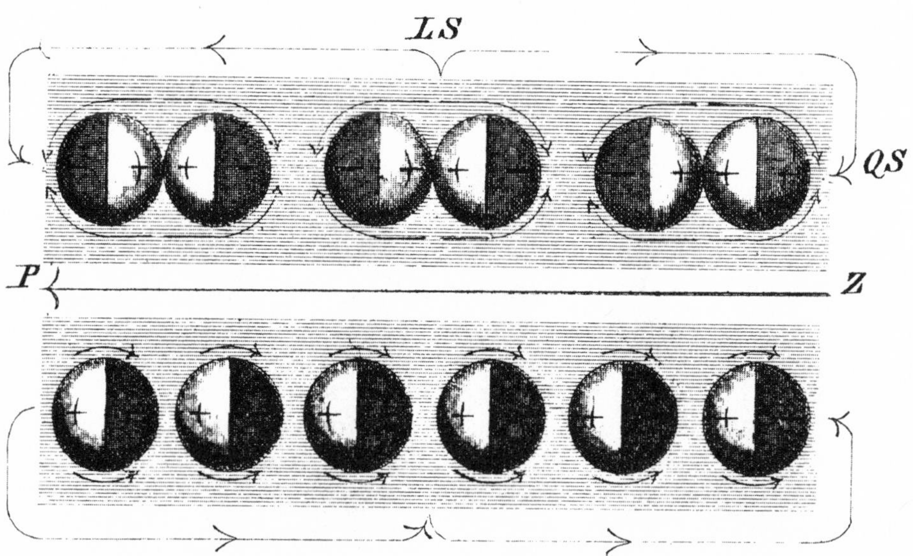

Dipolar Electrical Charge Distribution in a Nerve Tissue (19th Century)

ognize that particular types of experience may induce in our psyches "positive" or "negative" reactions that can store energy for later release, either gradual or cataclysmic, constructive or destructive, when triggered by some subsequent event. Frequent reference to such concepts can be found in the psychoanalytical literature; Freud himself describes how

> in mental functions something is to be distinguished—a charge of affect or sum of excitation—which possesses all the characteristics of a quantity (though we have no means of measuring it), which is capable of increase, diminution, displacement and discharge, and which is spread over the memory-traces of ideas somewhat as an electric charge is spread over the surface of a body.[12]

A positive example would be a sequence of compliments or fortunate occurrences that bring us to a state of crackling good humor; a negative example would be the day when everything goes wrong and our accumulated tension is discharged on some innocent friend who blunders into our emotional space.

Much as in physical electrodynamics, the presence of such emotional charge, with its attendant internal stress and pent-up energy, renders the consciousness susceptible to external forces that can affect its behavioral course. Common examples include the athlete or entertainer who becomes "charged up" for a game or performance; individual sensitivity to praise or criticism; the erection of psychological ramparts and defense mechanisms; crowd reaction to exhortation; or, in more extreme cases, emotional aberrations or illnesses of various types in which violent reactions can be triggered by relatively minor stimuli. Certain physical ailments with strong emotional overtones, such as epilepsy, hypertension, migraine, or allergies, may also illustrate aspects of this metaphor.

Consciousness Meters

Despite their conceptual identifications, none of these proposed consciousness metric quantities—distance, time, mass, or charge—offers any evident quantitative standard whereby it can be precisely measured. It may be that such gauges will ultimately be found within the human neurophysiological or neuropsychological experience-processing mechanisms, to the extent that those depend on unitary chemical and electrical effects. At present, however, such associations are well beyond our understanding of these basic processes, not least of all of the role of the brain as transducer and control center of the consciousness. The mathematician, Hermann Weyl, speculated on much the same dilemma, conceded the same inadequacy of comprehension, and affirmed the same reality topology:

> Between the physical processes which are released in the terminal organ of the nervous conductors in the central brain and the image which thereupon appears to the perceiving subject, there gapes a hiatus, an abyss which no realistic conception of the world can span. It is the transition from the world of being to the world of the appearing image or of consciousness.
>
> Here, we touch the enigmatic twofold nature of the ego, namely that I am both: on the one hand, a real individual which performs real psychical acts, the dark, striving and erring human being that is cast out into the world and its individual fate; on the other, light which beholds itself, intuitive vision, in whose consciousness, pregnant with images and endowed with meaning, the world opens up. Only in this "meeting" of consciousness and being both exist, the world and I.[13]

Even if these neurophysiological meters could be identified, it would probably be erroneous to attribute more than indicative functions to them. In the complex holistic processes of consciousness establishing its reality, these indicators are not likely to be any more central or causal than the feedback display circuits are to the operation of the REG or RMC machines.

Derived Properties

Physical analysis proceeds from its basic metric of distance, time, mass, and charge to define a host of useful secondary quantities, such as velocity, momentum, kinetic energy, and electric current, that are related to the primary metric in simple conceptual and algebraic terms. Similar arguments can be followed in the consciousness domain to erect a corresponding array of generalized properties useful for discussion and analysis of many individual and collective consciousness interactions. Of the many examples of this construction and their applications that could be detailed, let us mention just three.

"We would never doubt that the brain acts as a physico-chemical mechanism if treated as such; but for an understanding of psychic phenomena we would start from the fact that the human mind enters as object and subject into the scientific process of psychology."

WERNER HEISENBERG

Physical velocity is simply the rate of change of position per unit time. From our foregoing definitions, consciousness "velocity" can therefore be defined as the rate of acquisition of information per unit processing scan. Physical momentum is the product of mass and velocity—a weighting of velocity with the mass of the object. Thus, in consciousness space, if we weight the acquired information by its gravity or importance, the associated consciousness "momentum" assumes the role of the rate of acquisition of purposeful, consequential, or relevant information per scan.

Consciousness momentum is important in establishing a measure of the consciousness wavelength via the de Broglie relation (4.1). A consciousness that acquires purposeful or consequential information at a high rate should be characterized by a short wavelength; a long-wavelength consciousness employs many more scans to acquire the same amount of specific information with respect to a given goal. This by no means implies an intrinsic dullness of intellect or emotional insensitivity; rather, it relates to a broader mode of information acquisition that addresses a more diffuse pattern of goals, as epitomized, say, by the flight style of a butterfly, compared to that of a bumblebee. Such a style is consistent with the more free-flowing and aesthetic type of consciousness that, like impressionistic art, attends less to detail and pragmatics and more to holistic patterns and sensations.

Physical kinetic energy—one-half the product of mass and the square of velocity—is the energy possessed by an object by virtue of its motion. This kinetic energy, plus the potential energy the object possesses by virtue of its position in its gravitational and electromagnetic environment, is available for accomplishing change of its own state or of any portions of the environment it encounters. Comparably constructed consciousness "kinetic energy" can be similarly interpreted as the capacity for producing changes in the consciousness itself or in its environment. Thus, the higher the kinetic energy of consciousness operation, the greater the capacity to overcome environmental barriers, to influence other consciousnesses, or to change its own state.

"The true *motion* which lies at the basis of everything is the *motion of thought.* True *energy* is the *energy of consciousness.*"
P. D. OUSPENSKY

Carl Jung played with a similar set of analogies in his essay "On the Nature of the Psyche":

> The psychic intensities and their graduated differences point to quantitative processes which are inaccessible to direct observation and measurement. While psychological data are essentially qualitative, they also have a sort of latent physical energy, since psychic phenomena exhibit a certain quantitative aspect. Could these quantities be measured the psyche would be bound to appear as having motion in space, something to which the energy formula would be applicable. Therefore, since mass and energy are of the same nature, mass and velocity would be adequate concepts for characterizing the psyche so far as it has any observable effects in space: in other words, it must have an aspect under which it would appear as mass in motion.[14]

This metaphoric game could be pursued indefinitely to posit many more "mechanical" and "electrical" properties of consciousness—acceleration, angular momentum, electric current, magnetic dipoles, and others—all of which could serve to represent certain aspects of the consciousness/environment dialogue. Useful collective properties could be constructed as well: consciousness "thermodynamic" functions, such as temperature, density, pressure, and entropy; "transport" properties, such as conductivity, viscosity, or diffusivity; or "fluid dynamical" characteristics, such as shock waves, boundary layers, and turbulence. However, since none of these are essential to illustration of our basic approach, let us rather return to the primary metric and to the atomic-scale metaphors to develop the model in the most rudimentary terms.

6

CONSCIOUSNESS ATOMS

Our proposition, then, is to represent consciousness in terms of probability-of-experience waves, posed in the generalized consciousness coordinates, in much the same spirit as the probability-of-observation waves employed by quantum mechanics in physical space-time. In the absence of environmental constraints or interactions, these consciousness waves range freely over their own space-time domains, somewhat like the physical waves that ride the open ocean surface or those that propagate sound or light over large unbounded regions. But if a particular consciousness wave is confined to some sort of "container," or "potential well," representative of the environment in which that consciousness is immersed, characteristic patterns of standing waves, or eigenfunctions, will be established that represent the experiences of that consciousness in that situation (Fig. IV-2).

The dominant features of this environmental container are assumed to be associated with the living physical body upon which, for one mortal span, the particular consciousness wave has come to be centered. This basic physiological component is embellished by various proximate environmental details, both tangible and abstract, including physical surroundings, social context, other nearby consciousnesses, and, to a lesser degree, by more remote or global factors. Since all of these components are subject to change as the physical corpus matures, moves about, or interacts with other consciousnesses, or as the physical or social contexts change, and since the consciousness itself is continually changing, the eigenfunctions of experience are also constantly being altered. Nonetheless, at any point in this evolution they represent the tangible consequences of the consciousness/environment interaction, in much the same spirit as the atomic eigenfunctions of physical theory represent the observable properties of those systems. In other words, they define the consciousness atom.

In this view, any community of interacting personalities, rather than resembling a gas-kinetic ensemble of sharply localized and impenetrable particles, takes the form of a complex interplay of consciousness standing wave patterns, each centered in one of a corresponding myriad of mobile environmental containers. Every one of these wave-mechanical consciousness atoms is capable of interacting with its neighbors and with other aspects of its environment by all of the means available to the analogous physical wave systems, including wave-mechanical collisions, interatomic radiation, evanescent wave tunneling, or escape to free-wave status. Each of these forms of interaction bears its own metaphorical correspondence to various facets of normal and anomalous communication.

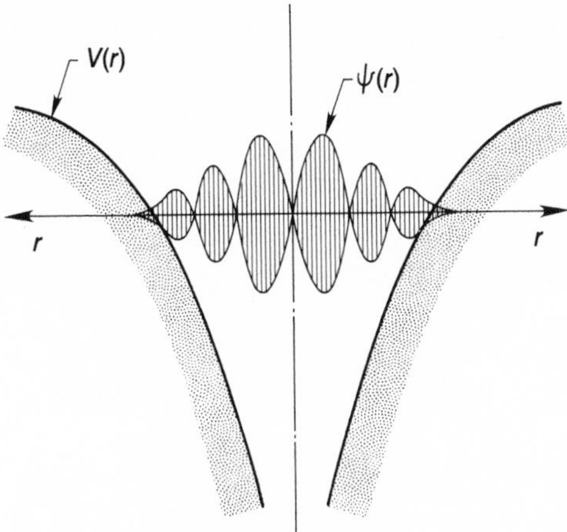

Fig. IV-2 *Consciousness Eigenfunctions* ψ(r) *in Environmental Well* V(r)

For example, borrowing the physical nomenclature, consciousness collisions can be catalogued as

a) *elastic,* wherein the two interacting partners, after some transitory distortions, return to their original configurations, with no permanent influence on the state of either participant;
b) *inelastic,* wherein the interaction permanently alters the state of one or both of the participants; or
c) *reactive,* wherein the basic character of the colliding system is altered, as in the formation of a composite consciousness "molecule" displaying quite different experiential characteristics from its atomic constituents.

Similarly, a consciousness "radiation" metaphor may be developed, wherein "photons" of information are transmitted from one individual to others throughout the community, to accomplish normal, or anomalous, forms of communication.

The process of evanescent wave penetration or "tunneling" may also represent various types of anomalous information acquisition, including remote perception and remote PK effects. As sketched in Fig. IV-3, standing wave patterns in finite potential wells are accompanied by some disturbances in the adjacent regions, even though the waves cannot propagate there. These evanescent waves can nevertheless convey limited information to and from other portions of the external environment, to a degree dependent on the details of the particular wave systems, and on the extent of the barriers separating them.

If any of the standing wave systems acquires sufficient energy to be elevated from cavity-bound to free-wave status, it may gain access to all consciousness space-time and interact with any other center in the configuration via that mode. Thus, this route could accommodate a variety of anomalies, including remote perception and remote man/machine interactions, as well as more extreme and controversial phenomena such as mystical union, out-of-body experiences, mediumship, and spiritual survival.

To pursue any of these metaphorical mechanisms in quan-

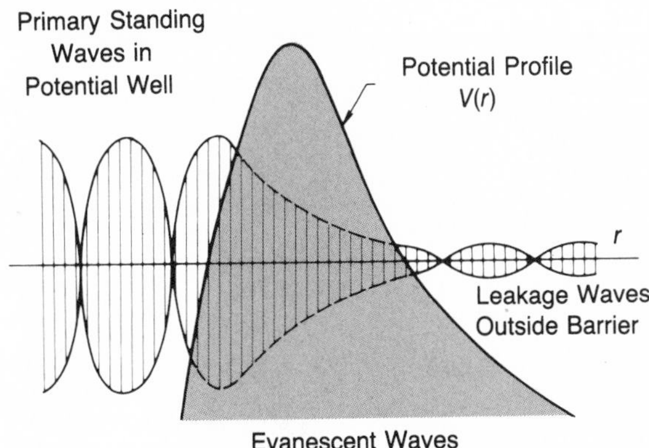

Fig. IV-3 *Evanescent Wave Penetration of Potential Barrier*

titative terms, the wave mechanics of the consciousness atoms must be formulated more explicitly. For this purpose, we employ the formalism of Schrödinger, embodied in Eq. (4.4), presuming that the conservation laws on which it is based and the mathematical operations it entails are valid in the consciousness domain. First, we must select a geometry in which the standing wave configurations of experience are to be constructed. Given its wide utility in representing a vast range of human experience on all scales from the subnuclear to the cosmological, its ubiquitous symbolism in all human times and cultures, and the inherently centered nature of the consciousness/environment model we are developing, spherical geometry would seem both analytically and intuitively appropriate as the frame of reference. (Schrödinger himself was wont to speak of "spheres of consciousness" in his philosophical writings.[15])

It then remains to specify the connotation to be applied to the geometrical "radius," "polar declination," and "azimuthal angle" in spherical consciousness space. Although the usual physical interpretations are clearly inadequate, they can again guide the generalization to the softer context appropriate to the excursions of consciousness. Thus, with reference to Fig. IV-4, we define the spherical consciousness coordinates, r, θ, ϕ, in the following impressionistic terms:

r: the radial coordinate, indicates the *range*, or depth of penetration, of the consciousness into its environment—the extent of its attention to it, or of its interaction with it.

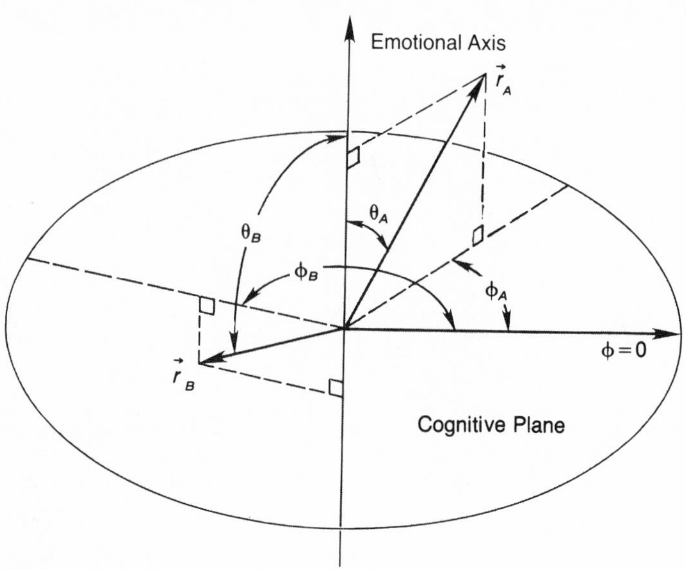

Fig. IV-4 *Consciousness Coordinate System*

θ: the angle of inclination of the range vector, $\vec{r}$, to the polar axis defines the *attitude* of the consciousness and thus specifies the emotional component of the interaction. If the inclination of $\vec{r}$ is upward from the equatorial plane ($0<\theta<90°$), the emotional attitude is positive (approach, attraction); if downward ($90°<\theta<180°$), the attitude is negative (avoidance, repulsion). The equatorial plane ($\theta = 90°$) denotes dispassionate cognition, and is henceforth termed the "cognitive plane."

ϕ: the angle specifying orientation of the component of $\vec{r}$ in the cognitive plane defines the cognitive *orientation*, point of view, or strategic perspective of the consciousness.

With these three coordinates, we can represent all variety of consciousness states quite parsimoniously. For instance, as sketched in Fig. IV-4, one may represent a strongly positive thrust of consciousness into its environment with cognitive orientation ϕ_A by vector $\vec{r}_A$; a mildly negative interaction with perspective ϕ_B by $\vec{r}_B$; and so on. Again, it might be noted that these apparently arbitrary soft coordinate specifications have some familiar linguistic associations; everyday speech frequently invokes similar geometrical terms to describe consciousness configurations. We speak of "spheres of influence," of "far-ranging" minds or thoughts, or of "penetrating" insight; we acknowledge the advantages of "positive attitude" or the detriments of "negativity"; and we ask "what's his angle?" or "what's the point?" to ascertain someone's perspective or orientation on a matter.

The environmental cavity that houses the consciousness waves and determines their particular standing wave patterns must also be represented in terms of these "soft" coordinates. For computational convenience, we assume that its profile depends only on the magnitude of $\vec{r}$ and not on either of the angular coordinates. More general forms could be accommodated with greater mathematical encumbrance, but little would be added to the basic conceptualization.

We can proceed from this point to derive mathematical expressions for the hierarchy of complex standing wave patterns allowed in any given confinement geometry. Like their physical counterparts, these patterns are indexed by three quantum numbers that serve on the mathematical side to specify the radial and angular nodal configurations of the standing waves, and on the conceptual side to quantify the probability-of-experience options. The principal quantum number, n, specifies the radial nodal structure and in the physical context is associated with the sequence of possible discrete energy levels of the atom. For our purposes, we interpret n as an index of similarly discrete options for the degree of investment, involvement, or attention of the consciousness, in either the cognitive or emotional sense, or both. As such, this property bears some similarity to various stratifications proposed in psychological, sociological, theological, and metaphysical models—"affect," "attentional energy,"

"The energetic activity considered in physics is the emotional intensity entertained in life."

ALFRED NORTH WHITEHEAD

"libidinal energy," "psychic energy," "rank," "hierarchy," "sanctity"—where in each case there is some attempt to quantize this elusive dimension. For example, several mystical traditions identify a hierarchy of energy centers, or "chakras," that underlie an individual's motivations or actions, or a sequence of degrees of enlightenment or spiritual growth. Various psychological theories suggest sequential stages of cognitive, emotional, or moral development, or hierarchies of psychological needs that progress from basic survival and primitive egocentric gratification toward more social concerns, ultimately reaching altruistic domains dominated by spiritual and religious values. Freud proposes a sequence of stages of ego development; Jung explores beyond the ego to deal successively with individual, group, and collective interactions requiring progressively more involvement; Maslow defines a hierarchy of personal psychological needs; Erikson identifies stages of emotional development; Piaget sets a scale of cognitive achievement; and Kohlberg similarly scales moral development. In social contexts, as well, we are accustomed to the establishment of discrete ranks—in the military, in business and industry, in

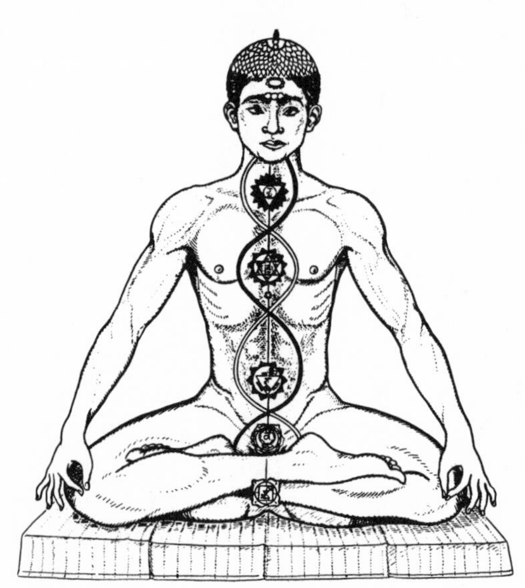

The Chakras

government, in religious and fraternal organizations and cults, in academia—that attempt to label in some crude way the nominal level of investment or achievement of the individual in the given environment. Without claiming any unification of these disparate stratifications, the quantum atomic metaphor provides symbolic representation of the common human tendency to quantize degrees of conscious or unconscious commitment or cognizance.

Whatever their psychological implications, these consciousness "energy" levels are subject to quantized change by mechanisms analogous to those of their physical counterparts, for example, by absorption or emission of information radiated from or to the environment, or by inelastic collision with other such consciousness atoms. Also in analogy to physical atomic structures, we may presume that beyond a certain level of consciousness energy, the wave pattern can no longer be contained by its environmental well and escapes to free-wave status.

The quantum numbers l and m specify the nodal configurations of the standing waves in the azimuthal and polar directions, respectively (Fig. IV-5). In physical applications, these are associated with the total angular momentum of the atomic system and the polar component thereof. (Note that such associations of gross mechanical properties with atomic-scale probability patterns are themselves heavily metaphorical, since the wave-mechanical model entails no moving particles.) In the consciousness metaphor, however, these quantum numbers relate to the arrangement of the probability-of-experience patterns in the cognitive-emotional space defined by our spherical coordinate system. Since probability patterns having large m remain close to the cognitive plane, while patterns of relatively small m are predominantly localized along the polar, or emotional, axis, we conclude that the ratio of m to l is indicative of the relative amounts of cognition and emotion prevailing in a given experiential interaction.

These metaphors can also accommodate the degree of complexity of the consciousness experience. For mathematical reasons, the angular quantum numbers l and m are constrained by the principle quantum number n to the following options:

$$l = 0, 1, 2, 3, \ldots, (n - 1) \tag{4.5}$$

$$m = 0, \pm 1, \pm 2, \ldots, \pm l\,. \tag{4.6}$$

Since l and m determine the number of angular nodes in the patterns, the more complex configurations can only be generated for the higher n's. Thus the analogy predicts that the more complex experiences of consciousness can only be attained with higher investments of consciousness energy. Again,

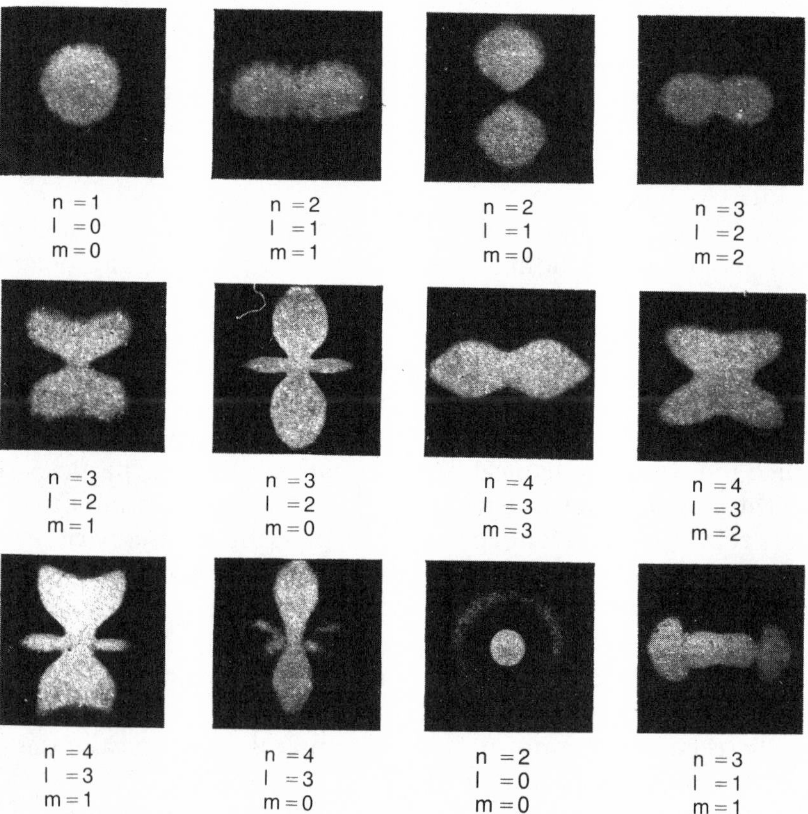

Fig. IV-5 *Shadow Illustrations of Electronic Probability Distributions in Hydrogenic Atoms for Given Quantum Numbers (after Born)*

the relative extent to which these complexities appear in the cognitive or emotional dimensions depends on the ratio of m to l.

One detail remains to complete this metaphorical consciousness atom—the property corresponding to the electron spin, s, mentioned earlier. In physical atomic systems this quantity does not contribute directly to the individual electron standing wave patterns, but manifests itself importantly in their interactions. Given its binary nature, its ubiquitous association with all atomic-scale systems, and its central role in the definition and application of the exclusion principle, we propose to associate this spin property in the consciousness metaphor with the fundamental "feminine/masculine" dichotomy, or the so-called "yin/yang" distinction of Taoist philosophy. More specifically, we shall use its two options to distinguish the passive/active, receptive/assertive, or enveloping/penetrating modes of consciousness interactions, whose balance is critical in interpersonal bonds.

Our consciousness atom has so far involved only a single "electron" in the individual environmental well. Extending the metaphor into "polyelectronic" structures would be tantamount to subdivision of the consciousness into identifiable components capable of interaction with one another, as well as with other consciousness systems. This multiplicity could prove useful for representation of many more complex features of consciousness activity, such as the ability of some individuals to segregate their patterns of attention and behavior into different sectors, or simultaneously to maintain multiple relationships with other consciousnesses on quite different bases. However, given the greater mathematical encumbrance of such models, we shall not pursue such polyelectronic formalisms here.

In summary, then, we have posed a quantum wave-mechanical model of the rudimentary consciousness atom, consisting of an array of spherical standing waves representing probability-of-experience patterns in a space defined by the intensity, attitude, and orientation of the consciousness in its interaction with its personal environment. The extent and configuration of these consciousness experience patterns are quantized, and are indexed in terms of the degree of involvement or attention, the complexity and balance of the emotional and cognitive components, and the assertiveness or receptiveness of the consciousness in the prevailing situation. With this model in hand, we can now explore specific classes of interaction.

Witches Frolic (19th Century)

7

CONSCIOUSNESS MOLECULES

One class of consciousness "collisions" posed in the previous chapter allowed for reactive interactions wherein the two colliding "atoms" would form a product "molecule" displaying categorically different experiential properties from those of its constituents. This particular branch of the metaphor has the capacity to represent many categories of normal and anomalous interactions and merits further development.

The quantum mechanical derivation of the characteristic wave structure of a physical molecule in terms of its constituent atoms is based heavily on the indistinguishability principle and the so-called "exchange energy" that follows from it. Via a method known as Heitler-London theory, it is possible to compute the energy levels of a diatomic molecule, E_{ab}, in terms of those of its two constituent atoms, in the form:

$$E_{ab} = E_a + E_b + E_x \,, \qquad (4.7)$$

where E_a and E_b represent the allowed energy states of the separate atoms, and E_x is an anomalous increment that derives primarily from the classically inexplicable exchange effect. Simply stated, when the wave patterns of the bonding electrons of the two parent atoms come into close interaction, they cannot be distinguished in any pragmatic sense. This loss of identity, when properly acknowledged in the quantum mechanical formalism, leads to an anomalous energy component in the bond.

As shown in Fig. IV-6, two possibilities obtain. If the spins of the two interacting electrons are the same, E_x is positive and the molecule is unstable and will not long cohere in this form. If the spins are opposite, E_x is negative and the molecular configuration is stable. The strength of the bond depends on the pattern of overlap of the two electronic wave functions within the composite potential well established by the two atomic nuclei and the electrons themselves.

The consciousness metaphor to this molecular bond has many useful interpretations. Our atomic model predicts that

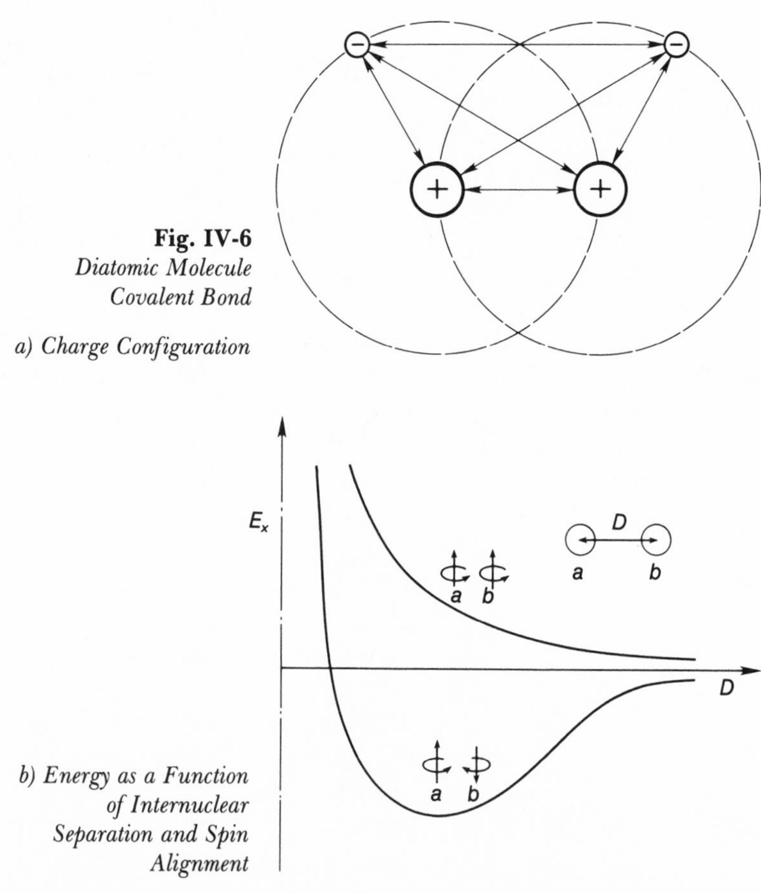

Fig. IV-6
Diatomic Molecule Covalent Bond

a) Charge Configuration

b) Energy as a Function of Internuclear Separation and Spin Alignment an individual consciousness immersed in a given environmental situation establishes a set of characteristic experiences. Let us call these ε_a. A second individual, exposed to the same situation, develops a different set of experiences, ε_b. However, if these two consciousnesses are strongly interacting, their experiential wave functions become intertwined in some resonant process, and their common environmental container is also modified by their mutual influence upon it. Thus the new pattern of standing waves established by the bonded pair in this modified environment embodies a set of experiences, ε_{ab}, that may be quite different from the simple sum of their atomic behaviors:

$$\varepsilon_{ab} \neq \varepsilon_a + \varepsilon_b \,. \tag{4.8}$$

In effect, the superposition of the atomic states must be augmented by a combination effect that includes all of the quantum wave-mechanical processes just described:

$$\varepsilon_{ab} = \varepsilon_a + \varepsilon_b + \varepsilon_x \,. \tag{4.9}$$

In any particulate representation, ε_x would be anomalous; in the wave-mechanical treatment, it is a *normal* consequence of the interaction. Thus it is ε_x that can represent our experimental anomalies.

"When simple cells become joined together, . . . they exhibit organizing forces in new directions which were impossible by any of the individual cells."

LUTHER BURBANK

For example, the remote perception results presented in Section III could be interpreted as a normal consequence of the interaction of a bonded percipient/agent pair within the experimental environment, properly different from the simple combination of experiences each would have had in the absence of the other's participation. Similarly, the man/machine anomalies of Section II could be interpreted in terms of a state of resonance between the operator and the device that yields a composite experiential output different from the calibration behavior obtained in the

Sea Gods (Roman Sarcophagus Cover)

absence of such coupling. Other examples, not addressed here, could also be posed, such as spiritual healing, mystical experiences, meditation, hypnosis, or simply the emotional reliance of one person on another, or of a child on a special toy. Many aspects of artistic and intellectual activity may involve similar mechanics of resonance and indistinguishability, and what we define as inspiration or creativity may stem from just such sacrifice of the atomic identity of the artist, scholar, or planner to the bond with his instrument or task.

The molecular metaphor need not be limited to binary interactions. A third individual or another impersonal element may be involved in the resonant system, much as in a triatomic molecule like H_2O or CO_2. Relevant examples in our experimental contexts include the co-operator protocols, the percipient/agent/target combination in the PRP experiments, and the role of an experimenter in enhancing results. Even more elaborate group interactions may be posed in analogy, say, to organic molecules, rings, and chains, or even to complex RNA and DNA structures. Here the anomalous group consciousness effects would correspond to the various modes of information and energy transfer and storage characteristic of such integral macromolecular structures, that are also inexplicable by any simple combination of the individual atomic behaviors. Alternatively, one could explore analogies to inorganic solid-state systems, such as crystals, metals, and semiconductors, where the organized or random behavior of a large number of interacting elements suppresses their individual characteristics within a diffuse, wave-like behavior of the ensemble. Examples of such resonant group situations are also apparent: the involvement of an audience with an artistic performance or sporting event; a political rally or mob hysteria; a religious service or meditation group; a military unit or orchestra.

Although this molecular metaphor has been presented in rather qualitative terms, at least one testable prediction and guideline for experimentation is evident. Namely, since the magnitude of the exchange effects that manifest as anomalies depends on the wave-mechanical resonance of the participating elements, stronger effects should occur among persons disposed toward sharing identity. Evident examples include the parent-child and sibling bonds, especially identical twins, and nongenetically based resonances between lovers, close friends or colleagues, teammates, and naturally empathetic persons in general.

Also implicit in the formalism is the prediction that the external environment, in the way it conditions the standing wave patterns, can facilitate or suppress the resonance and its associated anomalies. Again there are familiar examples: the romantic setting or comfortable office, the colorful excitement of a thronged sports arena or a military parade; or, on the negative side, the depressing effect of a hostile audience, a sterile workplace, or an unhappy home. There are strong indications that such ambience features can influence psychic experimentation as well, and it is for this reason that considerable care is taken to maintain comfortable and supportive conditions in our laboratory.

Before concluding this chapter, let us digress briefly to an issue glossed over in the preceding pages—the definition of the "consciousness" of a nonliving physical system. A few incidental comments may be relevant:

1) The distinction between living and nonliving systems, or between systems capable of "consciousness" or not, is becoming progressively more diffuse when approached from either the biological or physical perspective. Such biological entities as viruses, coacervates, and the life-forming DNA and RNA structures lie in a twilight zone between elaborate chemical complexes and vitally competent organisms.

2) The increasingly sophisticated microelectronic arrays employed in artificial intelligence, automated manufacturing, pattern recognition, robotics, and neural networks are approaching various forms of independent competence, judgment, and self-replicability.

3) The most fundamental premise of this model—that reality is established only in the interaction of a consciousness with its environment—need not be formally constrained to consciousness associated with a living system. Even in physical research, if I observe an experiment in progress, my reality is established by that observation. If I train some animal to observe that experiment, that animal's reality is established by its observations. If I replace the animal by an intelligent microelectronic device capable of observing and reacting to the experiment, in the same formal sense reality is established for that device by its observation process. (Whether either of the latter also constitutes a reality for *me* depends on my independent observation of the experiment, of the animal, or of the intelligent device.) Thus, in our consciousness metaphor, any functioning entity capable of generating, receiving, or utilizing information can qualify as a consciousness.

"The observer is never entirely replaced by instruments; for if he were, he could obviously obtain no knowledge whatsoever. . . . The most careful record, when not inspected, tells us nothing."
ERWIN SCHRÖDINGER

4) It is not unusual for operators of various types of machinery—automobiles, hand tools, pipe organs, or computers—to attribute anthropomorphic characteristics to their behavior. Similarly, many successful operators of the REG and RMC experiments testify informally to a sense that these physical devices also exhibit anthropomorphic qualities sufficient to engender quasi-personal relationships with them.

We shall presume, therefore, that within this metaphorical model, any device, situation, or task that takes on anthropomorphic aspects in the perception of a human consciousness and can be represented in terms of the same set of quantum numbers defining its intensity, emotion/cognition balance, and assertive/receptive characteristics acquires thereby the requisite properties of a consciousness wave function for purposes of interaction with its environment, including its perceiver.

Cover of Chalice Well near Glastonbury Tor

8

METAPHORICAL PRINCIPLES

If the proposed analogy to physical quantum mechanics has general validity for representing the interactions of consciousness with its environment, then the various quantum mechanical principles should also hold metaphorical relevance to human experience. Some of these have already been implicit in the proposition of consciousness atoms and molecules, but more specific associations can still be extracted.

The Indistinguishability Principle

As just discussed, the surrender of individual identity in a resonant bond with another consciousness or quasi-conscious physical entity may precipitate anomalous experience compared to the expected individual behavior. Again, the broader personal and social implications of this effect are empirically familiar: "It's bigger than both of us"; "Lost in each other"; "The whole is greater than the sum of its parts." Similar concepts underlie certain mystical traditions, such as the yogic "siddhis" achieved by the renunciation of personal identity, and the traditional tactics of transcendental or spiritual communion. However

"There is now strong reason to believe that *all* interaction forces in physics arise from the indistinguishability of the ultimate particles. Interaction has therefore a subjective origin."
ARTHUR EDDINGTON

expressed or practiced, the salient point for the design, implementation, and interpretation of experiments is that the blurring of individual identity in the intimate sharing of experience should facilitate anomalous results.

The Exclusion Principle

In the physical domain, quantum mechanics maintains that two electrons of the same quantum numbers cannot occupy the same space if their spins are also identical. Hence, molecular bonds are formed, the periodic table of elements is populated, and quantum statistical effects are determined by the interaction of electron pairs of opposing spins. In the consciousness domain, where we have associated spin direction with the passive/assertive or receptor/donor characteristics, the corresponding effect would be the tendency of bonds to form between two participants of opposite propensities of this sort. The most evident example is the ubiquitous female/male dyad, as acknowledged in numerous common expressions: "Every bird has its mate"; "It takes two to tango"; "Two's company, three's a crowd"; "Breathes there a man with soul so tough,

who can say two sexes are not enough." But the mechanics and implications of the consciousness exclusion principle are by no means restricted to feminine/masculine pairings in the physiological sense. The more fundamental requisite is that the yin and yang roles in the information exchange be balanced. If both partners are assertive, or both passive, little information can be shared and little bonding is likely to occur. Only if each role is filled can there be a stable molecular interaction.

These assertive and receptive roles need not be fixed; alternation may be equally effective. Indeed, in any form of dialogue between consciousness and its environment, both roles are inevitably performed by each partner; consciousness alternates between receiving information from and giving information to its environment. In a number of metaphysical traditions, this cyclical in/out dynamic is symbolized in terms of the physiological process of breathing, with its regular alternation of inspiration and expiration. Goethe invoked a very similar metaphor in the form of a maxim:

> Thinking and doing, doing and thinking, that is the sum of all wisdom. . . . Like breathing in and out both must always alternate.[16]

The Correspondence Principle

In the physical domain, quantum effects become evident when the dimensions of the object of interest and of its relevant environment are of the same or smaller scale than the de Broglie wavelengths of the wave functions that represent it; when the dimensions of the object and its environment are much larger than its associated wavelengths, classical continuum behavior obtains. Bohr's correspondence principle simply postulates a continuity of behavior over the transition between those two regimes. In our consciousness analogy, therefore, we anticipate that anomalous behavior will occur when the wavelengths of the consciousness wave functions have the same or larger scale than the dimensions of the environmental wells or contexts in which they are immersed, and will revert to a normal pattern when the wavelengths are much smaller than those dimensions.

Once again, the utility of this metaphor is limited by our present inability to specify more precisely the consciousness wavelengths or the dimensions of the environmental profiles that envelop them. Nevertheless, consistent with the consciousness de Broglie relation (4.1), we propose that the consciousness wavelength is at least qualitatively indicative of the scope or breadth of its involvement or, conversely, of the precision with which it addresses its context. More specifically, long-wavelengths are postulated to represent more free-flowing, holistic, generalized, aesthetic consciousness strategies, while short-wavelengths represent more precise, reductionistic, specialized, analytical styles. Similarly, the dimensions of the environmental well are also indicative of its generality or specificity: large dimensions imply a global or holistic context—the "big picture"; small dimensions imply a precisely defined, specialized context—the "immediate situation."

The metaphor then predicts that normal behavior will obtain when a consciousness function has sufficiently short wavelength for the context in which it must operate—when it can, as it were, define the characteristics of its environment with some precision, much as adequately short wavelength illumination is needed to cast a sharp shadow or form a sharp image of a given scale of object. In those situations where the wavelength of the consciousness function exceeds the dimensions of the context it is addressing, however, precise definitions will be replaced by various wave-mechanical processes such as diffraction, interference, or evanescent wave leakage, any of which may be interpreted as anomalous behavior.

"May it not be universally true that the concepts produced by the human mind, when formulated in a slightly vague form, are roughly valid for Reality, but that when extreme precision is aimed at, they become ideal forms whose real content tends to vanish away?"

LOUIS DE BROGLIE

It might be observed parenthetically that the traditional purpose of our formal and informal educational processes has been to develop the former class of consciousness interactions at the expense of the latter—to hone the consciousness to finer and finer precision, or smaller and smaller wavelengths—so that more and more specialized contexts may be addressed with increasing sharpness. Consciousnesses that have not been conditioned in this manner, such as very young children, primitive peoples, wild creatures, and those adults in our society who, for whatever reason, have resisted such training, do not automatically relate to their contexts with such precision, and indeed these groups seem more attuned to anomalous phenomena. In the pathological extreme, various forms of mental illness may cause serious inabilities to focus within the consensus frames of reference and can produce ineffable or incoherent impressions of reality.

The metaphor thus suggests a useful though vague criterion for facilitating the transition from ordinary to anomalous experience, namely that the consciousness involved become as freely ranging—that is, of as long a wavelength—as possible, compared to the prevailing context of the task. Again, we find some allusions to this requisite in various descriptions of states of consciousness associated with anomalous experiences, such as "unfocused," "diffuse," or "spaced-out," as opposed to "incisive," "sharp," or "uptight." For practical implementation of this criterion, it is interesting to speculate whether any of the various physiological oscillations, such as the brain wave patterns, the cardiac or pulmonary frequencies, or any other body rhythms, might in any way reflect the prevailing consciousness wavelengths, and thus be useful quantitative correlates for the phenomena.

The Uncertainty and Complementarity Principles

Heisenberg originally postulated a limitation on the precision of observation or localization of any pair of conjugate physical properties, such as momentum and position, or energy and time, in the mathematical form of Eq. (4.3). In pondering this principle, Bohr came to realize that its philosophical impact far exceeded its provincial applications in atomic-scale observations, and posed a broader version called the Comple-

Coat of Arms of Niels Bohr

mentarity Principle. This more general form addressed the interplay of alternative modes of perception and interpretation of any given event, of which the wave/particle duality itself was a leading example. Bohr's profound point was that although such modes may appear to be mutually exclusive, they actually should be regarded as complementing one another, so that via their superposition the full conceptual content of the event could be established. Subsequently, he extended this logic to even more categorical dichotomies, such as that between subject and object, or analysis and application. Nor was it necessary to restrict the case to physical logic alone; as Bohr himself put it:

> . . . we must, indeed, remember that the nature of our consciousness brings about a complementary relationship, in all domains of knowledge, between the analysis of a concept and its immediate application. . . . in associating the psychical and physical aspects of existence, we are concerned with a special relationship of complementarity which it is not possible thoroughly to understand by one-sided application either of physical or of psychological laws. . . . only a renunciation in this respect will enable us to comprehend . . . that harmony which is experienced as free will and analyzed in terms of causality.[17]

Heisenberg appears to have accepted such generalization. In a conversation with Bohr, he observed:

> Another argument . . . that is occasionally brought up in favor of an extension of quantum theory is the existence of human consciousness. There can be no doubt that "consciousness" does not occur in physics and chemistry, and I cannot see how it could possibly result from quantum mechanics. Yet any science that deals with living organisms must needs cover the phenomenon of consciousness, because consciousness, too, is part of reality.[18]

To which Bohr replied:

> The real problem is: How can that part of reality which begins with consciousness be combined with those parts that are treated in physics and chemistry? . . . Here we obviously have a genuine case of complementarity. . . .[19]

Elsewhere, Heisenberg admitted:

> We realize that the situation of complementarity is not confined to the atomic world alone; we meet it when we reflect about a decision and the motives for our decision or when we have the choice between enjoying music and analyzing its structure.[20]

Their colleague Wolfgang Pauli joined in this broader interpretation:

> On the one hand, the idea of complementarity in modern physics has demonstrated to us, in a new kind of synthesis, that the contradiction in the applications of old contrasting conceptions (such as particle and wave) is only apparent; on the other hand, the employability of old alchemical ideas in the psychology of Jung points to a deeper unity of psychical and physical occurrences. To us . . . the only acceptable point of view appears to be the one that recognizes *both* sides of reality—the quantitative and the qualitative, the physical and the psychical—as compatible with each other, and can embrace them simultaneously. . . . It would be most satisfactory of all if physics and psyche could be seen as complementary aspects of the same reality.[21]

For our purposes, we follow further along this philosophical vein to propose that many states of consciousness are inherently complementary to one another in much this same sense. In various pairs, these states embody the same type of mutual uncertainty or trade-off in precision of specification as do conjugate physical quantities. Without claim to completeness or uniqueness, a partial list of such "consciousness conjugates" could include:

analysis/synthesis
observation/participation
structure/function
goal/process
responsibility/independence
reasoning/intuition
doing/being

Note that in each case the two properties cited are not polar opposites, but are indeed complementary in the sense that the degree of each must be independently specified to localize the experience in that two-dimensional plane. But because of the uncertainty limits on localization, questions of balance or emphasis arise: highly structured reasoning obstructs free-flowing inspiration, but too much daydreaming precludes precise thought; overly reductionistic analysis interferes with holistic synthesis, while inordinate aesthetic demands can lead to chaos in implementation; too tight control, responsibility, or obligation can sap creativity and initiative, but unbridled independence of behavior becomes socially untenable; excessive activity or tranquillity can lead to states of anxiety or boredom, and so on.

Again, our language acknowledges the necessity for such compromise: "you can't have it both ways"; "he can't see the forest for the trees"; "he who would sacrifice his freedom for a little security, deserves neither freedom nor security";[22] "anything that liberates our spirit without giving us self-control is disastrous."[23]

In mathematical terms, the uncertainty principle derives from the intrinsic inability of wave mechanics to represent a sharply localized object without employing a broad range of wavelengths or, equivalently, from the inability of a narrow-band wave function to be confined to a small region. The corresponding technical argument for consciousness patterns is that if a given concept or task is to be addressed with precision, a broad range of consciousness wavelengths must be deployed to assess its full set of characteristics, and the state of the consciousness itself cannot be sharply specified. Conversely, if the consciousness is restricted to a single wavelength, i.e. to a "pure state," it cannot establish details about the task it is contemplating. Such pure states of monotonic attention are sought in various meditation techniques and mystical practices and are encountered in everyday experiences of "losing oneself" in a single activity, to the extent that it becomes impossible to describe precisely what one is thinking at the time.

From a more clinical perspective, the interest of some portions of psychology and physiology in lateralized hemispheric brain function may be another form of testimony to this consciousness uncertainty principle. To the extent that the functions predominantly carried by one hemisphere are complemented by functions of the other, the mechanics of their interactive balance may conform to this analogy. Conversely, extreme cases of hemispheric dominance may be accompanied by aberrations of perception or judgment in the inferior sector.

Beyond the limitation on precision of specification, the uncertainty principle may also define the most productive and fulfilling regimes of consciousness activity, namely where the doing and the being, or the analyzing and the synthesizing, are in some balance of focus. Again there are the common examples of the person who "lives his job," of the actor, dancer, athlete, or skilled artisan whose mechanical skills are complemented by subjective immersion in the role, resulting in a transcendent performance or product. Testimony from such genius or artistry commonly speaks of the necessity for dynamic balance between the skills and the immersion, the doing and the being, if the highest creativity is to be attained. Note also how we take greatest satisfaction in those technological devices in which structure and function are most harmoniously combined. A similar complementarity may well be requisite to the controlled generation of anomalous phenomena and could be a useful criterion for the design of appropriate experiments.

9

MAN/MACHINE MODELS

The opening chapter of this book spoke, in quite a different metaphor, of the strides of science alternating between its experiments and its theories and of the role of anomalies in correcting its course. With one foot based on the experimental data outlined in Sections II and III and the other probing tentatively forward in the theoretical direction just suggested, we are now poised to test our own steps in advancing comprehension of the role of consciousness in the establishment of physical reality. Although fully quantitative scientific dialogue is clearly beyond the capacity of our model at its present stage, some qualitative and even semiquantitative interpretation of the data, as well as some useful guidelines for experimental design, can be drawn from its concepts and vocabulary.

To summarize briefly, the overarching idea is this: if one retains the classical particulate image of human consciousness as sharply localized within the physiological frame, any possibilities of it directly influencing a physical device or process, or of acquiring information from targets remote in space and time, are inexplicable. If, however, consciousness is also allowed a wave nature, metaphorically akin to that attributed to atomic-scale physical systems, a variety of wave-mechanical processes become available by which marginal broadening of the options for reality can be achieved.

As a first application of this model, consider a human operator, seated before a device such as the REG or the RMC,

attempting to influence its output distribution by anomalous means. Represent the consciousness of that operator as a quantum mechanical wave pattern expressed in the conceptual coordinates of range, attitude, and orientation defined in Section IV-6 and displaying, in the prevailing environmental context, a set of experiential standing waves indexed in terms of their intensity (*n*), complexity (*l*), ratio of cognition to emotion (*m*), and receptive/active "spin" index (*s*). Similarly, represent the essence of the experimental device by a similar set of wave patterns, expressed in the same conceptual coordinates and indexed by the same set of quantum numbers, indicative of its functional and aesthetic character as perceived by the operator.

It is the prediction of our model that if the wave pattern of the operator is not resonant, in the formal wave-mechanical sense, with that of the device, there can be little anomalous interaction between them. Rather, they will retain their atomic identities, with no discernible change in the normal behavior of either. If, however, some wave-mechanical resonance is established between them, molecular experiential patterns can arise whose observable characteristics differ significantly from the simple sum of their individual behaviors. The basic experimental strategy, therefore, should be to devise practical means for enhancing such wave-mechanical resonance.

The degree of this resonance depends on details of the superposition of the wave patterns of the operator and of the machine in the prevailing experimental environment. Thus, the desired tuning could be addressed in three ways, or some combination thereof: 1) adjustment of the operator's wave pattern to match that perceived for the device; 2) configuration of the device wave pattern to match that of the particular operator; or 3) adjustment of the environmental profile within which the operator and machine interact during the experiment, to optimize the superposition of their individual wave patterns.

To develop any of these options, it must be recognized that the perception of the device in wave-mechanical terms may vary from operator to operator, depending on his technical sophistication, psychological state, and experimental strategy. In some cases, its salient features may derive from the character of the feedback by which it shares its information with the operator, so that the quantum numbers n, l, m, and s that characterize its wave pattern refer primarily to the intensity of that feedback, its complexity, the balance between its cognitive and emotional impact, and its passivity/activity, respectively, as perceived by the operator. Other aspects of the machine, such as its overall appearance and scale, the noise it makes when functioning, or even some imparted anthropomorphic characteristics, may also contribute, but for most technically unsophisticated operators it seems unlikely that any of the internal mechanisms will be relevant.

However the device characteristics are established, to proceed with the first tuning option, the operator would attempt, by some psychological strategy or intuitive subconscious scanning, to identify and maintain his own wave pattern that best resonates with that of the device as he perceives it. For example, he might endeavor to adjust his own wavelength to some resonance with the machine by altering his momentum (Eq.

[4.2]), either via a change in his consciousness mass (gravity/levity) or via a change in his velocity (rate of experience per processing scan).

In the second tuning option, the perceptible character of the machine's operation could be altered by changing some parameters of its information-processing mechanisms—by setting different numbers of samples, sampling rates, or counting modes, or by employing feedback of different prominence, intensity, or complexity—in a search for the combination that best resonates with the consciousness state of the operator. Variation in the run or series length might also alter the character of the experiment as perceived by the operator.

For the third option, some aspects of the experimental protocol, the ambience of the experimental area, the pace of the data acquisition, or the role or style of the laboratory staff could be tuned to the expressed or perceived preferences of the operator. For example, the sensitivity of some operators to the instructed/volitional or automatic/manual secondary parameters suggests that these aspects of the protocol can affect their resonance with the machine.

Perhaps the single most critical aspect of any of these tuning processes, however, is the correlation of the spins of the operator and the device. The quantum mechanical formalism predicts that only if these spins are opposite can a productive resonance prevail; if they are the same, the interaction is repulsive, and anomalous effects opposite to the intended direction may be expected. To tune the spins, the operator might first assess whether the device seems to be active or passive in its relationship to him and then adjust his own style accordingly. Alternatively, he could maintain a fixed spin characteristic and explore various operational modes that emphasize the perceived activity or passivity of the device. For example, the automatic mode of machine operation, wherein trials are initiated and results presented via the feedback at regular intervals, out of direct control of the operator, might be perceived as an assertive behavior of the device to which the operator must respond passively. In contrast, the manual mode, wherein the operator initiates each trial at his pleasure, might be perceived to place the operator in the assertive role and the device in the receptive. A similar difference of perception might attend the operator's freedom to choose his direction of effort, vis-à-vis having to accept the random instruction of the machine. When optional feedback modes or other subjective strategies are employed for this purpose, the process may be further complicated by successive inversions of the spin roles of the operator and device during a trial, a run, a session, or a series, so that both phasing and frequency become important in retaining the resonance.

Correlation of the experimental data with the various tasks, devices, and modes employed and comparison with the informal reports of operator techniques are at least qualitatively consistent with aspects of the bonded-system strategy just proposed. For example, the achievement of most operators is quantitatively, and often qualitatively, different for the two directions of effort, PK^+ and PK^-. Given the complete symmetry in the technical operation of the equipment and in the protocols, this disparity must presumably be related to subjective factors in the operator's approach to the task, such as conscious or unconscious psychological nuances he associates with "going high" or "going low" (or right or left) that may manifest themselves as differences in his wave patterns when approaching the two tasks. Beyond this asymmetry, the characteristic specificity of individual operator signatures bespeaks a complex and elusive tuning process having many degrees of freedom, some technical, some psychological.

The model suggests a number of other possible correlations that are currently under experimental study, or are planned for future investigation. For example, the programs of remote and precognitive PK experiments address the sensitivity of the man/machine resonance to spatial or temporal separations and may help sort out the importance of such physical parameters in these bonds. Experiments employing more than one operator on a device at a given time, or more than one device for a given operator, may illuminate the wave-mechanical combination of influences and efforts. In the design and interpretation of these and other attractive experimental possibilities, the generic concept of operator and machine achieving a resonant wave-mechanical bond, via a variety of technical and psychological tuning strategies, and thereby generating significant statistical variations from the normal behavior of the unbonded man/machine system, continues to seem pragmatically and philosophically useful.

The experimental data and informal operator reports are equally consistent with a somewhat different metaphorical representation of the anomalies in terms of a tunneling or barrier penetration process of the operator's consciousness. In a sense, the difference between a chance distribution of binary events and the slightly displaced distributions achieved in the PK efforts may be regarded as a marginal increase in the information content of the bit sequence that is in some way attributable to the consciousness of the operator. In any particulate representation of consciousness, no mechanism for such transfer of information can be identified within the conditions and protocol of the experiment; all are precluded by the total encapsulation of the consciousness within the physical body. In the wave-mechanical representation, however, some diffusion of information is possible, utilizing the evanescent standing waves of consciousness that extend beyond the dimensions of its environmental well.

Our model predicts that this diffusion of information should be favored by keeping the environmental barrier between the consciousness of the operator and the external system as small as possible and by employing consciousness patterns of long wavelength and large amplitude. In more subjective terms, these transcribe into a "proximity," in consciousness coordinates, to the PK task ("feeling close"); a reduction of the body-centeredness of the consciousness; and a relaxation of high-precision analytical thought processes in favor of more broadly ranging, but nonetheless intense, diffuse-focus modes, respectively, all of which seem consistent with the reported strategies of many of our successful operators.

Whichever of these wave-mechanical options are applied, the model and the experiments agree that under certain circumstances human consciousness can interact with physical systems to broaden their observable behavior beyond chance expectation. They also concur on some qualitative conditions for such broadening. More quantitative agreement of experiment and theory must await better definition of the consciousness coordinates, the consciousness quantum numbers, and the consciousness quantum itself.

Abraxas (Gnostic Gem)

10

PRP INTERPRETATION

The precognitive remote perception phenomena also lend themselves to interpretation via similar quantum mechanical metaphors. Resonant bond models can be applied in various ways, depending on the roles presumed for the percipient, agent, and target. For example, the anomalous information acquisition may be regarded as a proper consequence of a resonance between the percipient and agent, with the target characteristics entering as components of the environmental profile that condition the experiential wave patterns of the bonded pair. To the extent that this bonded pattern differs from the separate atomic patterns of the participants, it can accommodate both the anomalous information acquired by the percipient about the target as well as any influence of the percipient on the agent's pattern of attention to the target.

Alternatively, the percipient may be regarded as resonating with the target rather than with the agent, leaving the agent relegated to the role of conditioning the molecular environmental pattern. This version, which seems consistent with those experiments employing percipient/agent pairs who are total strangers, or no agent at all, first requires assignment of a wave function and quantum numbers to the target scene, in much the same spirit as for the PK machines. The task for the percipient's consciousness is then to tune its own wave pattern, by any strategy it finds effective, until some resonance with the unknown target is detected, and thereby to infer some aspects of its wave-mechanical, and thence its physical, characteristics. Yet a third option would be to treat all three participants—percipient, agent, and target—on an equivalent basis in analogy to the triatomic molecule mentioned in Section IV-7.

Quantum mechanical tunneling or free-wave propagation metaphors may also be serviceable for interpretation of the PRP results. If the tunneling model obtains, three interrelated optimizations should be sought in the percipient's strategy: 1) the maximization of the amplitude and range of the penetrating evanescent wave; 2) its localization on the desired remote target; and 3) the discrimination of valid information acquired from the target over background noise. In the theoretical formalism, the first aspect is again favored by minimization of the environmental barrier opposing the wave and by maximization of the consciousness wavelength and amplitude. The localization and discrimination steps suggest a vectoring and tuning process, perhaps akin to that employed in aircraft traffic control or sonar sounding, where subtle resonances between the percipient's wave pattern and that of the target are detected. Alternatively, the target or agent may be regarded as constituting a local distortion in the overall environmental profile that functions as a beacon or marker for the percipient's searching evanescent wave.

The reports of many of our PRP participants are consistent with such tunneling mechanisms:

> "First, I try to shut down whatever I've been doing or thinking about, then I try to envision a blank movie screen on which I'm about to see a film starring the agent that I've never seen before. Then I wait to see what comes up on the screen. Usually, though, it's not a sharp picture but more like a sequence of dream-like impressions."
>
> "Most of my perceptions are performed first thing after I awake—before any logical traffic enters my mind. After defining the purpose of my search, I lie very still, and just wait for images to flow in."

In essence, it appears that, by various psychological strategies, percipients first attempt to suspend the analytical, short-wavelength activity of their usual cognitive processes that by their nature are sharply localized and incapable of much penetration. This leaves available their longer-wavelength, impressionistic awareness to diffuse outward and scan the remote environment. To enhance this diffusion, these percipients may also endeavor to reduce the environmental barrier surrounding their normal tightly centered realities by relaxing the degree of attachment of their consciousness to its physical locus.

In extreme cases, a percipient may report that his consciousness seems to have been totally liberated from its center to roam freely in space and time, which would correspond, in the metaphor, to escape to the free-wave state. In this state, the consciousness could access remote locations as an outward propagating wave, reflections of which could return information about those locations. This process may be akin to the so-called "out-of-body" or "near-death" experiences where individuals claim to view their own bodies and surroundings from an external vantage point.

Of all of the features of the PRP data, the precognitive aspects are perhaps the most striking and challenging, and also the most difficult to accommodate in any theoretical formalism. Nonetheless, this model also offers some rationalization of this feature, either in terms of a penetration by the percipient's consciousness through an intervening temporal barrier or in terms of a free-wave propagation in the time dimension. Recall that the basic consciousness waves, in their unbounded form, have access to all of the consciousness space-time net. It is only through the imposition of environmental constraints that the space and time localizations of the consciousness standing waves of experience are established. If these constraints are re-posed in some fashion, both space and time may, in effect, be redefined. That is, rather than forming its experiences in the "here and now," consciousness may choose to sample the "there and then." In more sophisticated quantum mechanical formalisms cast in relativistic coordinates, there is little mathematical distinction between spatial and temporal behavior, so

Remote Perception in Cyrano de Bergerac

that any metaphoric mechanisms invoked to explain acquisition of information remote in distance would equally well apply to information remote in time. Finally, of course, we might recall Einstein's and Eddington's reminders that the concepts of space and time are themselves constructions of consciousness.

The choice between the resonant bond, barrier penetration, or free-wave metaphors for the interpretation and strategy of remote perception experiments is still unclear. Indeed, these concepts embody many similar conceptual and pragmatic features that may make the distinction more semantic than practical. The essential point is that various nonclassical effects of quantum wave mechanics bear useful similarities to these anomalous experiences of human consciousness.

11

BROADER BASIS

Numerous other data on anomalous consciousness-related phenomena, acquired under various degrees of experimental rigor and precision over the past century, could also be addressed via this wave-mechanical model. While we shall not attempt this here, the interested reader may enjoy pondering applications of the metaphor to these diverse research results, or to yet more controversial claims of levitations and poltergeists; apparitions and teleportations; natural resource and archaeological dowsing; remote surveillance and crime detection; psychic healing and diagnosis; and the ultimate propositions of reincarnation and the survival of bodily death. Turning it to less controversial terrain, the model can also serve to represent many aspects of "normal" human behavior that strain particulate interpretation, such as love, humor, creativity, inspiration, intuition, or friendship, or any other experiences where interactive resonances or bonds with persons, pets, tools, or tasks are major factors.

Another potentially fruitful class of application of the metaphor is to collective systems such as crowds or societies, where quantum statistical formalisms may be generalized to predict the most likely circumstances for anomalous behavior. Recall that for some atomic-scale ensembles, the Pauli exclusion principle alters the recipes by which individual particles can assemble in proximity to one another, and thereby distorts their statistical distributions from the classical expectation. Actually, two such categories of quantum statistical behavior can prevail: the so-called Fermi-Dirac statistics applicable to particles of half-integer spins, and the Bose-Einstein statistics for particles of zero or integer spin. The extent to which either system departs from classical behavior turns out to depend on the physical ratio $n/(mT)^{3/2}$, where m is the mass of the individual particle, T the prevailing temperature of the ensemble, and n the density in which it is assembled. Thus, deviations from classical behavior are favored by low particle mass, low ensemble temperature, and high density or concentration.

One could propose that collective interactions of consciousness might involve similar quantum-statistical behavior and thus appear anomalous in comparison to conventional expectations. For example, strictly transposed, this metaphor would predict that a concentrated array of consciousness interactions pursued under attitudes of low mass (levity, lightheartedness) in an ambience of low temperature (tranquil, cool) would have

The Holy Grail Appearing to the Knights of the Round Table

greater likelihood of displaying departures from classical statistical behavior. In fact, this recipe is an appropriate description of a variety of social configurations that purport to elicit anomalous effects, such as religious services, séances, and various communal retreats. It is also consistent, in quite a different sense, with the ambience prevailing in our laboratory: equipment and protocols facilitating concentrated attention on dense ensembles of digital information; playful operator attitudes; and a low-key environment.

In introducing this model, we acknowledged that the choice of quantum mechanics as the theoretical framework to be generalized was rather arbitrary, and that other forms of physical theory, such as electricity and magnetism, statistical thermodynamics, special and general relativity, or even classical Newtonian mechanics, could similarly be appropriated for metaphorical representation of consciousness processes. Certainly the concepts of the consciousness coordinates and metric transcend the particular equations for which they were needed here and could equally well serve these alternative formalisms. Beyond that, it was our basic premise that *any* physical theory only represents the interaction of consciousness with its physical environment and thus reflects the processes by which consciousness perceives, orders, and represents the information it acquires in that interaction. Consequently, Maxwell's equations, the laws of thermodynamics, and even Newton's Second Law inevitably testify to the nature of the consciousness that devises and applies them and, therefore, should also provide useful metaphors for representation of that consciousness.

"It is the theory which decides what we can observe."
ALBERT EINSTEIN

"Theory like mist on eyeglasses. Obscure facts."
CHARLIE CHAN

One particular physical domain, adjacent to the quantum-atomic regime adapted here and possibly worthy of appropriation for future extension of this model, is that of nuclear and subnuclear phenomena. In particular, this could lead to better understanding and specification of the environmental profiles that condition the consciousness standing wave patterns, in the following sense. The physical potential well experienced by atomic electrons is established by the array of charge localized in the tightly packed ensemble of nuclear particles and it is these configurations that are represented by the various formalisms of theoretical nuclear physics. The substructures of the nuclear particles are also modeled by extensions of those techniques into the domain of high-energy physics. Many of these representations, like our consciousness metaphor, employ quasi-atomic conceptualization and formalism. The proposition could thus be made that the consciousness environmental wells are similarly established by various internal arrangements of consciousness "nucleons" maintained by a hierarchy of nuclear "forces" analogous to those of the physical models. One could further speculate that these consciousness nucleons may bear some relationship to certain intense and deeply set psychological factors, such as Jung's "archetypes." This extension of our model would require an effort of at least the scale already expended and will not be developed here. Nevertheless, some attempt to represent human consciousness as it turns inward toward its center, rather than outward toward its environment, could be a worthwhile next step.

"The microphysical world of the atom exhibits certain features whose affinities with the psychic have impressed themselves even on the physicists. Here, it would seem, is at least a suggestion of how the psychic process could be 'reconstructed' in another medium, in that, namely, of the microphysics of matter."

CARL JUNG

In an even broader sense, metaphorical representations of consciousness need not be restricted to physical or even to general scientific vocabulary and concepts, but may appropriate their analogies from any province the human mind has attempted to organize or correlate. For it is precisely through such processes of organization and correlation that the essential characteristics of that mind are revealed. In the words of William James:

> Out of this aboriginal sensible muchness attention carves out objects, which conception then names and identifies forever—in the sky "constellations," on the earth "beach," "sea," "cliff," "bushes," "grass." Out of time we cut "days" and "nights," "summers" and "winters." We say *what* each part of the sensible continuum is, and all these abstracted *whats* are concepts.
>
> *The intellectual life of man consists almost wholly in his substitution of a conceptual order for the perceptual order in which his experience originally comes.*[24]

Indeed, it could be contended that although our metaphor has been cast in physical concepts and vocabulary, the ultimate product has been a psychological model. This is not accidental, for in the largest sense, both physics and psychology address the question of how man perceives and represents his world. Certainly there are many evocative similarities in the evolution of psychology from its empirical and classical phases through the once radical propositions of James, Freud, and Jung, on to its contemporary specializations, and the development of physics from its Newtonian era through quantum mechanics into its current esoteric proliferations. Einstein recognized the epistemological parallel quite clearly:

> Body and soul are not two different things, but only two ways of perceiving the same thing. Similarly, physics and psychology are only different attempts to link our experiences together by way of systematic thought.[25]

From any perspective, the essential point remains: what consciousness experiences, it attempts to comprehend; to comprehend, it must order; to order, it must name; and in the naming, it *creates* its experience. No more profound or enduring testimony to this truth can be cited than the simple account of Genesis 2:19 (King James):

> And out of the ground the Lord God formed every beast of the field, and every fowl of the air; and brought them unto Adam to see what he would call them: and whatsoever Adam called every living creature, that was the name thereof.

This timeless allegory acquires even greater contemporary relevance if the "animals" are allowed to include black holes and quarks, space and time, waves and particles, all other aspects of the environment, and, of course, consciousness itself.

SECTION FIVE

The Vectors Deflected

If we wish to give philosophic expression to the profound connection between thought and action in all fields of human endeavor, particularly in science, we shall undoubtedly have to seek its sources in the unfathomable depths of the human soul. Perhaps philosophers might call it "love" in a very general sense—that force which directs all our actions, which is the source of all our delights and all our pursuits. Indissolubly linked with thought and with action, love is their common mainspring and, hence, their common bond. The engineers of the future have an essential part to play in cementing this bond.

—LOUIS DE BROGLIE, *New Perspectives in Physics*[1]

Madonna Nebula

1

MOTIVATIONS AND IMPLICATIONS

Seldom do we present a technical seminar, entertain visitors to the laboratory, or engage in discussions with our professional colleagues that some question does not eventually arise concerning our motivation in pursuing these studies. Occasionally, such queries are cast in quite direct, even blunt terms, such as "What are nice folks like you doing in a field like this?" or "I'm surprised to find someone of your background interested in such topics." But more often the challenge is presented more obliquely, with averted gaze, a shuffling of feet, or a nervous cough that betrays a deeper concern of the questioner for the personal implications that might attend the answer. However clothed, the question is fair; involvement in so challenging and contentious a topic should be based on clear personal evaluation of its short- and long-term implications.

Our response usually acknowledges the same triad of motivations attributed to ancient mystical man and persisting through all subsequent scholarly attention to the metaphysical dimensions of human experience: the pragmatic, the intellectual, and the spiritual. To be sure, we tend to cast these in contemporary and sophisticated terms, but doubtless their roots trace to the same archetypal aspirations that impelled our academic ancestors. Actually, these three interwoven purposes can be said to characterize all of the best engineering enterprises of any kind, in their attempts to provide technological devices, systems, and processes combining effective function, enhanced comprehension, and aesthetic satisfaction.

Thus, as engineers, we are first concerned with the practical implications of the observed phenomena and of the theoretical model required to represent them. From a cautionary standpoint, we must consider the possibility that a broad range of devices and systems generally similar to those employed in our research may likewise be responsive to the conscious or unconscious intentions of their human operators, to an extent that could complicate their various technological applications. From a more positive perspective, this same responsiveness may open paths to new generations of engineering equipment and processes wherein human consciousness can play an active, integral role. At the least, better comprehension of such subtle capabilities of human consciousness may abet the development of progressively more sophisticated devices that can mimic those same subtle functions. In all cases, there is an engineering responsibility to study, to implement, and to control, overlaid with a concern for greater compatibility of man and his machines.

Also as engineers, we are committed to the collection, correlation, and interpretation of basic knowledge. To pursue that purpose in this particular topic, our scientific two-step requires the acquisition of large amounts of reliable empirical data on the anomalous effects of interest, accompanied by a broadening of the theoretical paradigm to accommodate the consciousness-related parameters that bear on the observed phenomena. With either foot we tread on somewhat strange ground for traditional scientific research, and any progress made may have wider implications for the growth of knowledge in many other domains.

"Man must cling to his faith that the incomprehensible is comprehensible, else he would cease to investigate."

GOETHE

Important and attractive as these pragmatic and intellectual motivations may be, we also acknowledge the role of the third dimension—the spiritual implications, in the fullest sense of the term—in imbuing this engineering research with yet broader relevance and excitement. Here we refer not to any quasi-religious zeal or dogmatic conviction, but to the same desire to improve the quality of life, and to the same quiet philosophical reverence for the majesty of the cosmos and man's role in it that have been acknowledged by most creative scholars throughout history. In this category, for example, we ponder whether incontrovertible scientific demonstration of the ability of consciousness to influence its physical reality—if such can ever be attained—would substantially alter individual and collective perception of the human state, its value system, and behavior patterns. Also in this dimension, we endorse de Broglie's vision of the "engineers of the future" at both the philosophical and tactical levels. Careful application of scientific knowledge and rigor of method, within a permeating atmosphere of " 'love' in a very general sense," is clearly a noble and powerful plan for relating thought to action in any technical arena. But the formula appears particularly pertinent to this topic in its specific efficacy for realization of the phenomena themselves, for it is precisely through the bonds of consciousness established between human operators, or with their technical devices and processes, that the anomalous data seem to emerge. Thus, the spiritual component participates in a very pragmatic sense: selfless investment of self can affect physical reality.

> **"For in every act of love and will—and in the long run they are both present in each genuine act—we mold ourselves and our world simultaneously."**
>
> ROLLO MAY

Our statements of motivation frequently precipitate a second challenge: given the array of experimental results and the predictions of the wave mechanical model, what *are* the scientific and philosophical implications, or, again in very blunt terms, "so what?" One mode of response to this equally legitimate and complicated query is to return to the confluence of conceptual and topical vectors outlined in Section I and inquire whether they may be affected in any way by our experimental and theoretical results—whether they may, as it were, be refracted by their passage through this research event, to proceed henceforth along somewhat modified paths. The bulk of this final section will explore these individual deflections and attempt to weave their new configuration into a structural network that can support a more comprehensive representation of reality. For this purpose, we shall permute the order somewhat from their earlier presentation, to permit systematic progression from the pragmatic through the intellectual and spiritual domains.

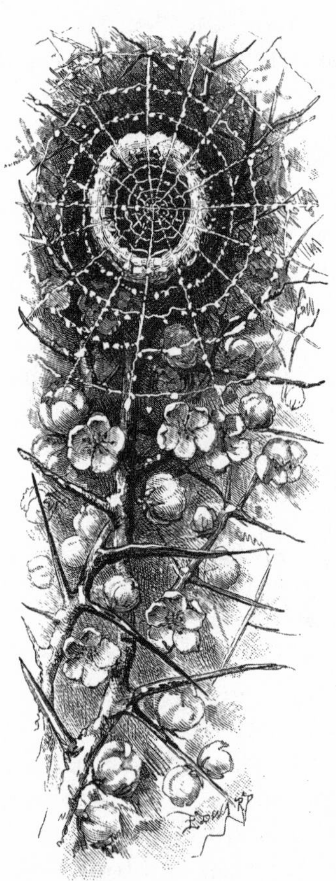

2

MAN/MACHINE MARGINS

Earlier we raised various chimeras of possible "high-tech" malfunctions attributable to anomalous operator interactions with microelectronic gear or other low-level information-processing devices and systems. The experimental data, along with certain aspects of the theoretical model, have reinforced such concern, at least over the range of equipment and parameters they have addressed. In essence, we have found that a variety of devices based upon random or pseudorandom physical processes can display marginal aberrations in their output distributions that correlate with individual operator intentions. While the scale and character of these alterations vary substantially from operator to operator, their individual statistical replicability leaves little doubt about the reality of the effect. Over large blocks of data, the deviations can accumulate to highly significant levels that would far exceed the tolerance of many delicate information-processing devices commonly employed in major data management systems, control machinery, and diagnostic apparatus. Thus there arise legitimate questions about the functional integrity of such classes of equipment in the absence of adequate protection from, or compensation for, such operator-related influences. There also arises an array of fascinating possibilities for constructive utilization of such effects in a variety of engineering applications.

In the former category, imagine the cockpit of a modern combat aircraft or of a future high-performance commercial transport in which a vast array of indicators and controls, many involving microelectronic elements akin to those employed in our research, present the crew with a surfeit of technical information and functional tasks related to the myriad of interdependent subsystems that govern the operation of the craft. In the light of our experimental results, can we confidently presume that all of these highly integrated devices and systems will continue to function in nominal fashion, totally independent of the conscious or subconscious intentions of the operators? Or is it possible that the same sort of marginal aberrations observed in our basic experiments might also arise here, on a scale capable of affecting the overall performance of the aircraft?

Consider similar scenarios in an air-traffic control tower, in an ICBM silo, in a nuclear power plant, in a neurosurgical operating theater, in a major investment data-processing facility, in an automated industrial factory, or in any other application where human and artificial intelligence must operate reliably in close concert, under sensitive or intense conditions. What margins of uncertainty can be anticipated or tolerated

here? What strategies of insulation, limitation, or compensation can be devised? How far may we trust our "intelligent" or "expert" machines in such situations? Do the basic research results justify some *applied* research to assess such issues more directly? Certainly the continued drive toward more and more delicate and sophisticated systems in complete ignorance of such assessment seems irresponsible.

Approaching the matter from a more positive perspective, it seems equally reasonable to inquire about possible constructive applications of these effects. As a superficial example, had our data-processing algorithms been equipped with an additional logic stage that provided some output signal when the cumulative deviation from any experiment exceeded a preset threshold, it would have been possible to trigger some palpable effect or process at that point, such as opening a garage door, turning off a television set, or letting in the cat. Given the statistical nature of the effects, the interoperator variability, and our inadequate present understanding of the roles of operator motivation and purpose, such direct applications should probably be deferred, or only pursued with great conservatism. Possibly more attainable, and certainly more consequential, would be the development of more sophisticated, efficient, and creative man/machine systems based on better comprehension of the subtle capabilities of consciousness in bonded-system interactions.

A major uncertainty in all of this speculation is the extent to which anomalous effects observed in experiments with a few specific engineering devices, performed by a small group of operators, in a particular laboratory environment, can reasonably be extrapolated to implicate broader ranges of technical equipment, wider segments of the population, and less structured work environments. Clearly, much more extensive experimentation and much more comprehensive and quantitative theory will be needed to support or discredit such potentialities. Nonetheless, the possibility that certain man/machine interactions may require generalization of traditional mechanistic models to include consciousness as an active, synergistic component of the overall system has legitimately been raised.

The results of the experiments in precognitive remote perception also suggest a number of potential applications of engineering relevance, some of which are in fact already in provisional practice. As mentioned earlier, similar techniques have been employed, albeit in supporting roles, in such diverse areas as national security and law enforcement, archaeological searches, natural-resource prospecting, and medical diagnosis. Just as in our basic research, the emphasis in most of these applications has shifted from credibility demonstration to refinement of protocols, participant strategies, and analysis techniques to enhance their efficacy. The relevance of our efforts to such applications has been primarily in the analysis category, where digitized evaluation techniques like those described in Section III have proven useful in rendering results into systematic and quantitative forms.

Of much wider and more fundamental long-term importance to engineering practice, however, is the possibility that some redefinition of the basic concept of *information,* in the practical sense of the term, may be required. Just as operator-specific marginal variations in output distributions may arise in sensitive man/machine interactions, PRP indicates that similar volitionally induced margins of information may be extracted from classically occluded sources. In cases where the classical prediction is a random or noise distribution, significant positive, or negative, increments of information may appear through the intervention of human consciousness. In situations where the classical distribution already entails some finite information content, that too may be broadened to higher or

"In the sense of traditional physics, information is neither matter nor energy. Rather, the concept of information brings into play the two older antipoles of matter—namely, form and consciousness. . . . One must take 'probability' and 'information' as objective and, at the same time, as subject-related concepts."

CARL FRIEDRICH VON WEIZSÄCKER

lower values. In short, any of the conventional sensory modes of information acquisition may conceivably be enhanced by increments anomalously acquired via the consciousness bonds.

Paradoxical as this consciousness-induced alteration of the information distributions of physical devices and processes may at first appear, it is quite consistent with the geometry of reality proposed in Section IV-2. Namely, if reality is established only by the interpenetration of consciousness and its environment, then the sole expression of that reality—information—must be the child of both parents and hence reflect the influence of both. The situation is aptly illustrated by the legendary response of a famous major-league umpire when asked how he made the marginal judgments when calling pitches behind home plate. "Well," he said, "it ain't all that hard: some of them's balls and some of them's strikes, and some of them ain't anything 'til I call them."

Thus, to summarize the broader implications of our results for engineering and technology, we propose that the established deterministic view of the behavior of information-processing devices and systems may need to be generalized to allow a margin of output behavior that is in some way reflective of individual operators' conscious or subconscious intentions in the prevailing or perceived interaction. The practical importance of this margin will depend on the sensitivity of the physical system, its application, the surrounding circumstances, and the degree of resonance between it and its human operator. When the combination of these factors is negligible, classical behavior will obtain; when their overlap is propitious, anomalous behavior may ensue.

3

PARAPSYCHOLOGICAL PROGRESS

As contemporary psychic research continues on its lonely and controversial path, its primary hope for scientific credibility remains the establishment of much greater statistical replicability of its experimental data base. Over the past few years, major strides have been taken toward restructuring parapsychological research to focus on this issue. Via more careful combinations of experimental equipment and designs, more incisive data-processing strategies, and more extensive interlaboratory collaboration, the data base has gradually grown to more convincing status. For example, over the past decade there has been a notable increase in the sophistication of the experimental protocols and equipment, permitting greater accuracy and rapidity of data acquisition, recording, and processing. Closer attention to calibration procedures and fail-safe controls has considerably reduced the possibilities of technical artifacts or hidden malfunctions of the experimental devices, while on-line automatic data collection has minimized operator and experimenter involvement in raw data recording. With these facilities, data can be collected in sufficiently large blocks, under sufficiently uniform conditions, that unequivocal trends can emerge from the background random noise. Crucial to this strategy has been the use of simple experimental designs and differential data collection strategies, such as the tripolar protocols, that ensure separation of consciousness-related effects from all other possible sources.

There is also a growing realization that data acquired, processed, and interpreted primarily on an operator-specific basis display finer-grained aspects of the phenomena and better lend themselves to more systematic correlation with psychological and physiological measures. Within these individual operator data bases, intention becomes the primary variable, with technical and protocol options playing secondary roles of varying importance, and the particular nature of the device apparently being least important of all. In many PK experiments, and in ESP studies as well, anomalous effects that may be quite marginal on any individual trial basis are found to compound with statistical regularity to significant levels as the data bases grow and the search for major aberrations has been subordinated to more systematic accumulation of such marginal increments. Results are now sought across broad ranges of participant personalities, physical devices, and experimental protocols, with less emphasis on so-called "gifted subjects," now that the replicability of ordinary operator data bases has proven more than sufficient to allow credible correlation and interpretation along various empirical and theoretical lines.

"Psychic research presents a challenge that science can no longer avoid. . . . It can be an important element in the long-sought formula for enriching human awareness, reconstructing society, and generally aiding nature in the great work of evolution."

EDGAR D. MITCHELL

The recent efforts have also attempted more systematically to relate empirical data and experimental designs to theoretical models of various degrees of sophistication. This provincial version of the scientific two-step has been inhibited in the past by the limitations of many of the postulated models, but as improved experimental designs serve to discredit, circumscribe, or modify a number of the inadequate theories, the surviving superior models can indicate directions for more substantial empirical progress.

On the political front, this same era has seen parapsychological research alter its traditional defensive posture to engage in more constructive dialogues, and at times even productive collaboration, with scholars outside its immediate community, including some of the major critics of the field. Of particular interest have been the cooperative efforts to develop credible methods of so-called "meta-analysis," whereby results of similar experiments at various laboratories may be pooled. Such studies have demonstrated some commonality of marginal yields across substantial ranges of superficially disparate studies.

Looking to the future, the need continues for more rigorous experimental designs and methods of analysis, more disciplined restraint in the extrapolation of effects from inadequately small data bases, and the conception and development of more powerful theoretical models. Paramount in all of this is the requirement for yet larger and better-balanced data bases to facilitate discrimination among the myriad of factors possibly

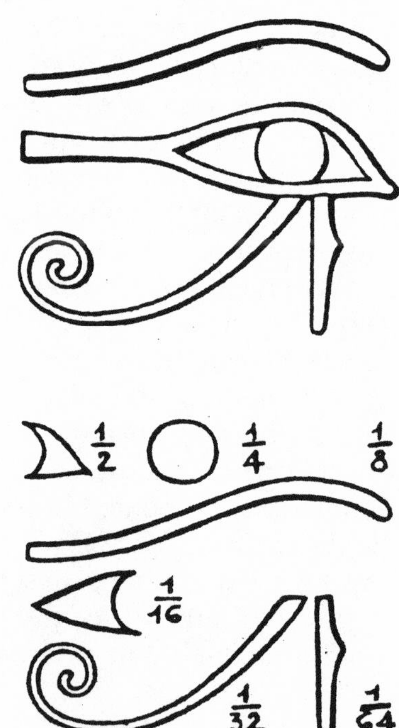

The Eye of Horus: Holistic Imagery versus Quantitative Analysis

contributing to the phenomena. Closely related is the desirability of greater interlaboratory cooperation, manifested in more compatible experimental designs, nomenclature, and data-handling techniques. Also suggested is further integration of the operators into the experimental and theoretical teams.

As this more sophisticated parapsychological strategy yields more replicable and credible new data, the phenomena addressed should become more widely acknowledged. The scope of this field will then broaden to provide additional interfaces with other established branches of science, engineering, and numerous less technical disciplines, that in turn will further strengthen the campaign. In this sense, parapsychology may soon find itself at the historical and epistemological hub of a new generation of truly interdisciplinary scholarship.

4

CRITICAL COLLOQUY

The traditional marketplaces of scientific scholarship are the meetings of its professional societies and the pages of its archival journals, where proponents of particular experimental results or theoretical ideas may set out their wares for confirmation or criticism by their peers. The proper role of the critic in these forums is to mount responsible challenges to the proposed evidence and ideas, so that sound information and concepts may be strengthened by contest and fallible material winnowed out. To fulfill this important function productively, the criticism must be both informed and fair, and it must deal in relevant facts and specific argument rather than resorting to categorical rejection, dogmatic recalcitrance, or guilt by association. And, of course, it should be scientifically altruistic, rather than self-serving. Regrettably, the critical commentaries on many classes of anomalous phenomena reported throughout scientific history have not always abided by these rules, and hence have required at least as careful sifting as the targets of their criticism to eliminate the dross of incompetence, special interest, and scientific bigotry. Only then has the valid residue been able to resume its proper participation in the research dialogue.

The most common criticisms about the conduct, results, and interpretation of contemporary research into anomalous consciousness-related phenomena were discussed in Section I-5, along with their relevance to the design and implementation

A. absolute Gravity. B. Conatus against absolute Gravity. C. partial Gravity.
D. comparative Gravity E horizontal, or good Sence. F. Wit. G. comparative Levity.
or Coxcomb. H. partial Levity, or pert Fool. I. absolute Levity, or Stark Fool.

Satire on Newton's Gravitation (18th Century)

of our own program. Reexamination of these issues in the light of the specific results of that program can suggest a more constructive course for future excursions of the critical community into this topic.

In assessing possible naïveté or fraud in the conduct of this or any other research, no sweeping assurances from either side are likely to be as persuasive as incisive independent examination of the equipment, protocols, and results by responsible and objective peers. Better yet would be formal replications of the results by other laboratories or by the critics themselves. To these ends, complete summaries of all experimental data bases, technical equipment, protocols, and data-processing strategies should be made available, for whatever use any critic or colleague may choose to make of them. It is then incumbent upon the critics to avail themselves of those facts, and to base their comments solely and logically upon them. We believe that the results outlined in this book and detailed in our journal articles and specialized technical reports obviate to a considerable degree the critical complaints regarding poor experimental replicability and correlation. However, it should be emphasized that this statistical replicability resides primarily in the operator-specific experimentation. Indeed, this operator-specific replicability is a major finding of our research and seems indicative of the fundamental nature of the phenomena.

"Even scholars of audacious spirit and fine instinct can be obstructed in the interpretation of facts by philosophical prejudices. The prejudice . . . consists in the faith that facts by themselves can and should yield scientific knowledge without free conceptual construction."

ALBERT EINSTEIN

The critical contention that deviations from chance expectation are only marginal in magnitude is in fact sustained by this research at the level of small individual groups of data, but the intrinsic statistical replicability of individual operator performance allows these marginal effects to accumulate to high levels of significance as the data base grows. While we cannot claim to have systematically addressed the issue of the elusiveness of the effects under critical or hostile scrutiny, we can report that significant data have been generated in our laboratory by operators of initially skeptical dispositions.

Criticisms concerning the absence of viable theoretical models and the inconsistency of the purported phenomena with the prevailing scientific paradigm have been addressed by a model that concedes the need for major conceptual expansion to accommodate consciousness-related effects of this type. However, the metaphors and formalisms on which this generalized model is based are all drawn from established scientific concepts; only the domain of their application has been broadened.

In sum, we believe that many of the traditional criticisms of consciousness-related anomalies research have been considerably weakened by the results of this and similar contemporary programs, and that the critical community must seek firmer ground on which to pursue its proper role. Specifically, the burden of negative proof should now turn to identification of specific and consequential technical flaws in the experimental equipment, design, or interpretation; the generation of countervening data using these or similar experimental facilities;

Captain Bligh's Drawing of the "Anomalous" Echidna

the posing of testable alternative hypotheses for interpretation of the data; or the construction and operation of competitive experiments of comparable sophistication. Such approaches could reenergize the iterative dialogue between advocates and critics until the reality of the phenomena is established, one way or the other.

"For if those who hold that there must be a physical basis for everything hold that these mystical views are nonsense, we may ask—What then is the physical basis of nonsense? . . . In a world of aether and electrons we might perhaps encounter *nonsense*; we could not encounter *damned nonsense*."

ARTHUR EDDINGTON

Throughout this dialogue, both sides need to recognize the immense professional and personal investments in the negative status quo that properly prevails throughout the majority of the scientific establishment. Science cannot allow itself to be unsettled by every apparently anomalous effect that crosses its horizon, particularly when major tenets of its established paradigm would be threatened, and it has every right to demand extraordinary proof of extraordinary claims. Similarly, personal resistance to fundamental changes in established belief systems can be a healthy restraint to overreaction of the individual psyche to superficial aberrant stimuli. Nonetheless, that same science, and that same personal psyche, must have the openness of mind, the humility in the face of empirical evidence, and the flexibility of spirit to accommodate to new information when such is incontrovertibly presented, for therein lies the only route for their vital growth and maturation.

5

STATISTICAL SENSITIVITIES

The evidence for consciousness-related anomalies presented in this book rests on systematic statistical deviations from chance expectation of large strings of data acquired under different states of operator intention, as epitomized, for example, by the REG and RMC cumulative deviation signatures. The similarity of such signatures obtained on substantially different devices argues against the effect being microscopically mechanistic—such as a physical distortion of the electron current pattern in the microelectronic noise diode, or some unspecified force exerted on the balls in the RMC machine. Rather, a more holistic alteration of the overall information content of these systems seems to be occurring, manifesting itself in distorted output distributions. But it is less clear whether these observed aberrations trace directly to corresponding biases in the unit probabilities of the individual bits of information processed by the system or to some modification of the statistical recipe whereby they are compounded.

Familiar examples of the bit-level class of bias would be an unfair coin or a loaded die that alters the unit probabilities from strictly 1/2 or 1/6, respectively, to slightly different values, causing correspondingly displaced statistical patterns of results. Similarly, in the PK experiments, one could postulate that the role of the consciousness has been to alter slightly the basic binary probabilities, and that this alteration then compounds through traditional statistical mechanics to yield the distorted output distributions. If this were indeed the mechanism at work, we would infer from the terminal results that the unit binary probabilities have typically been so biased over a range of only a few tenths of one percent.

"In the universe great acts are made up of small deeds."
LAO TSU

The second, or combinatorial, class of distortion has some precedent in the domain of quantum statistics mentioned very briefly in Section IV-4. In our case, the proposition would be that if consciousness interactions are subject to an exclusion principle, large assemblies of repeated interactions could conform to quantum statistics, rather than to classical statistics, leading to significant deviations from classical expectations.

Whether or not either of these mechanisms actually applies to the laboratory experiments, some involvement of consciousness in the prevailing statistical behavior of the effects seems inescapable and, in a sense, even reasonable. In a broad view, all perceptual, cognitive, and emotional processes of consciousness may be regarded as intrinsically statistical in character. Whenever we are exposed to any sensory stimuli, a variety of elaborate mental organizational programs of comparison, discrimination, association, and prediction, all involving probabilistic logic, are employed to convert these stimuli to meaningful information. As we reason our way through problems, alternatives are automatically assessed on a probabilistic basis. Even

I Ching

our emotional anticipations are scaled by the perceived likelihood of future threats, challenges, or pleasures. This innate statistical sense doubtless underlies the formal statistical concepts and methods devised by consciousness to represent its experiences in quantitative terms. In other words, statistics, like all other theoretical concepts, is a subjective matrix imposed by consciousness to help it sort stimuli into information, and thus may be as much reflective of the consciousness itself as of the effects it is sorting.

In the apple-tree analogy employed in Section I-8 to illustrate rudimentary statistical concepts, it was noted that the characteristic breadth or variance of the statistical distributions about their mean values could subsume all manner of potentially influential parameters, such as location, season, weather, and apple type, that for whatever reasons could not be deterministically included in that model. In our experiments, it appears that the consciousness of the operator qualifies as such a potentially influential parameter and hence must either be specified as a fixed index of the individual data bases or subsumed as an uncontrolled variable contributing to the variance of the overall results. This, of course, is precisely the distinction between the individual versus the collective cumulative deviation graphical presentations.

Since operator intention also appears to be a primary variable, a second statistical discrimination is necessary to make sense out of the signatures. Namely, the PK^+, PK^-, and BL traces need to be displayed separately; if they are lumped together, most operator results look like chance behavior. For some operators, further specification of subordinate parameters, such as volitional/instructed or automatic/manual modes, is also needed to display the effects clearly.

One hypothetical interpretation of these experimental results is that consciousness imposes some form of anomalous segregation of the output data during its acquisition. For example, the terminal series score distributions displayed in Fig. II-11 show significant shifts in the means and increases in the variances of the PK^+ and PK^- distributions, but a decrease in

"One must frankly admit that even at the human level life only advances hesitatingly under the influence of great numbers and the play of chance. . . . The work of creation presents itself to our experience as a *process* whose laws it is the business of science to investigate, leaving to philosophy the task of discerning in the phenomenon the place and influence of *intention*."

PIERRE TEILHARD DE CHARDIN

the variance of the baseline distribution. Yet, when all of these are recombined in a balanced mixture, the total distribution essentially reconstructs the chance Gaussian. Thus, one possible postulate is that the function exercised by consciousness in this situation is preferentially to select from the grand chance distribution of potentially available scores those subsets that support its teleological purpose—to obtain high, low, or baseline values—without altering the overall distribution.

From all of this emerges the intriguing possibility that what we denote as chance behavior, in any context, rather than deriving from some inviolate mechanistic behavior of a deterministic physical world, is actually some immense subsumption of a broader distribution of potentialities reflective of all possible resonances and intentions of consciousness with respect to the system or process in question. In this sense, the concept of "margins of reality" posed specifically for these experimental results, would be escalated to apply to any form of probabilistic experience. Arthur Eddington, the British Astronomer Royal, described the situation in only slightly different terms:

> It seems that we must attribute to the mind power not only to decide the behaviour of atoms individually but to affect systematically large groups—in fact to tamper with the odds on atomic behaviour. . . . Unless it belies its name, probability can be modified in ways which ordinary physical entities would not admit of. There can be no unique probability attached to any event or behaviour;

> we can only speak of "probability in the light of certain given information," and the probability alters according to the extent of the information.[2]

However the matter is phrased, we now begin to converge on a generic response to the keynote question posed in the Preamble regarding the nature and magnitude of the role of consciousness in the establishment of physical reality. Namely, each consciousness has its own private margin of reality to play with, consistent with its own innate statistical sense and information-processing style, and scaled by its ability to achieve a resonance with its environment.

6

QUANTUM CONSEQUENCES

One of the strongest persuasions toward our particular theoretical approach was the widespread interest of many of the pioneers of modern physics in the potential relevance of their quantum mechanics to other domains of human experience, including anomalies of the sort we are addressing here. Most of the members of this elite community displayed a refreshing openness to possibilities of psychic phenomena in general, not rejecting the topic out of hand, but reserving their judgments until more credible data could be assembled. Certainly they shared, in varying degrees and perspectives, the sense that physical theory would not be complete until it incorporated consciousness in some form. The research outlined in this book modestly offers two related responses to these challenges. First, it submits several substantial empirical data bases that could be useful to contemporary quantum theorists in readdressing the consciousness issue. Second, it proposes one possible approach to modeling the mechanics of consciousness, via an expansion of the quantum mechanical metaphor that acknowledges its source as well as its object.

Assessment of the relevance of our experimental and theoretical efforts to the present stance and future course of quantum physics involves a hierarchy of interwoven scientific and philosophical considerations, each of which has its own profound implications. Quantum mechanics, in any form, clearly replaces strict deterministic logic by probabilistic statements about physical properties and processes. The Copenhagen interpretation goes beyond this abstract statistical stance to restrict the relevance of the mechanics to *observed* properties and processes; it deals only in probability-of-observation theory. As such, this interpretation implicitly concedes a role for consciousness in the specification of physical reality. Our experiments and model support a statistical view of reality in general, and the Copenhagen stance in particular, and may have some utility for their refinement. The essence of the experimental results is that the involvement of human consciousness with a probabilistic physical system or process avails statistical ranges of options for its tangible output that can be correlated with the individual and with his state of intention, as well as with the prevailing technical conditions. As suggested in the previous chapter, any form of statistical behavior already implies some subjective component in the experience; the demonstrated specific correlations with individuals and their intentions is evidence that this component is not only active, in the Copenhagen sense, but is also causal, in a volitional sense.

This volitional causality is not explicitly acknowledged in the traditional or contemporary formulations of quantum theory. Whether these data are sufficiently persuasive, or the metaphoric model sufficiently useful, to support such further reinterpretation of so successful a formalism is now at issue. On the one hand, it seems unlikely that most of the highly successful applications of quantum theory to calculation of atomic-scale properties will be substantially altered by inclusion of an explicit consciousness component in the mechanics. On the other hand, this component conceivably may be the missing ingredient needed to resolve certain persistent paradoxes, such as those relating to the so-called "Einstein-Podolsky-Rosen (EPR) paradox," that appear to demonstrate anomalous action-at-a-distance between separated parts of a previously bonded system.[3,4] In a broader philosophical view, however, one could propose that all of the apparently arbitrary quantum caveats from which these successful formalisms originally devolved

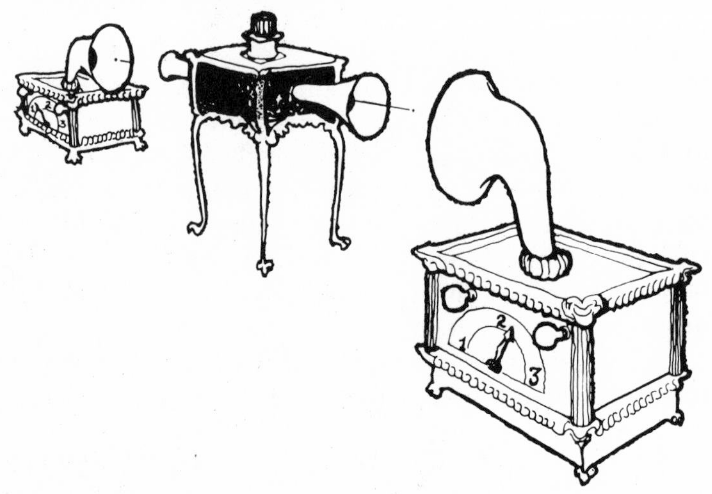

EPR Experiment (Schematic)

may themselves now be logically explicable from this more generalized, consciousness-incorporating perspective. In particular, if such primary postulates as the wave/particle complementarity, the basic quantizations, the uncertainty and exclusion principles, and the observability criterion, are regarded as archetypal propensities of all consciousness/environment interactions, rather than as peculiar requisites of the province of atomic physics, then all of the subsequent mechanics emerge as more universally grounded and therefore more relevant to any realm of human experience. In a sense, this is the converse proposition to our appropriation of quantum mechanics as a metaphor for consciousness. Now we invert the logic to suggest that it is consciousness that is the ultimate metaphor for quantum mechanics, or for any other theoretical model, for that matter.

In pursuing this philosophical approach, we might remember that the correspondence principle assures us that most quantum physical anomalies become insignificantly marginal as the scale of the system in question approaches macroscopic dimensions. It is only on the atomic scale that most quantum wave-mechanical effects are striking. Similarly, the extraction of consciousness-related anomalies from the output characteristics of our macroscopic experiments requires very large data bases because the marginal effects are proportionately so tiny. In interactions with physical atomic domains, the consciousness margins of reality may become comparable in scale with the basic processes themselves, and thereby acquire much greater relative importance. In fact, they may just *be* the basic processes *in toto*, and all atomic reality may be established by these mechanisms alone.

A similar suspicion may have underlaid Schrödinger's radical proposition:

> . . . the "real world around us" and "we ourselves," i.e. our minds, are made up of the same building material, the two consist of the same bricks, as it were, only arranged in a different order—sense perceptions, memory images, imagination, thought. It needs, of course, some reflexion, but one easily falls in with the fact that matter is composed of these elements and nothing else. Moreover, imagination and thought take an increasingly important part (as against crude sense-perception), as science, knowledge of nature, progresses.[5]
>
> The only possible inference from these . . . facts is, I think, that I—I in the widest meaning of the word, that is to say, every conscious mind that has ever said or felt "I"—am the person, if any, who controls the "motion of the atoms," according to the Laws of Nature.[6]

In a similar tone, Eddington speculated:

> We may look forward with undiminished enthusiasm to learning in the coming years what lies in the atomic nucleus—even though we suspect that it is hidden there by ourselves.[7]

Other scientists and philosophers have pondered whether atomic structure may be characterized by its own intrinsic form of consciousness. By the definition of consciousness proposed in Section IV-7, atoms and molecules would certainly qualify, for they have the capacity to exchange information with each other and with their environment and to react to these in some quasi-intelligent fashion. But any attempt to distinguish between such inherent atomic consciousness and one imposed by our own interactions with these entities is probably unnecessary and unproductive, since it is we who ultimately establish the concepts of either form, and since *our* reality transpires only in those interactions.

Eddington's allusion to the role of the mind in the atomic nucleus may be particularly prophetic, for as science continues to probe to smaller and smaller scales, the consciousness margin may become more and more dominant. At the close of his career, Heisenberg conceded that the basic question of the composition of nuclear particles may be illegitimate and will

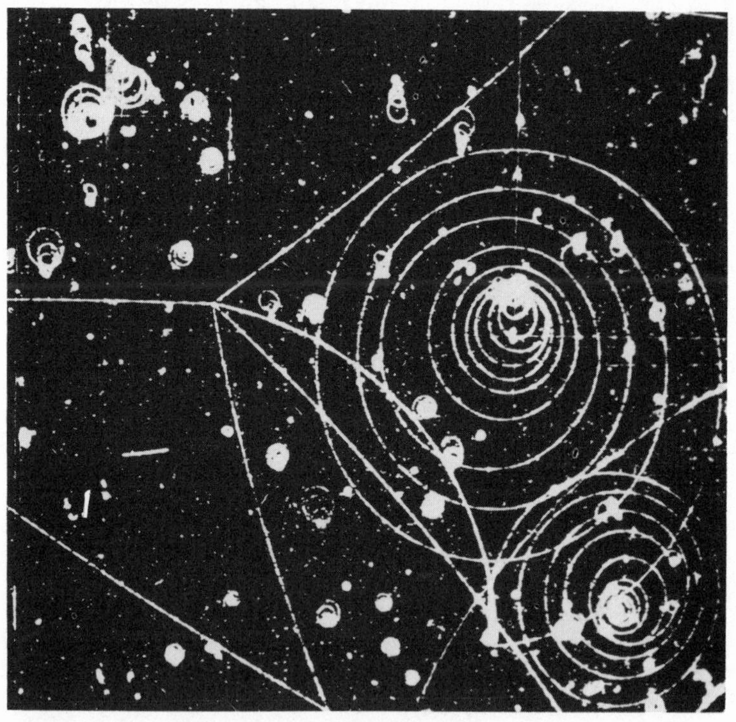

Cloud Chamber Photograph of Nuclear Particle Trajectories

Spiral Nebula in Canes Venatici

yield sensible answers only if those particles, or their components, can be observed without investment of energies of preparation or observation comparable to their rest masses.[8] Putting it more bluntly, the high-energy accelerators may not be identifying nuclear structure; they may be creating it. And the products of this creation, the "quarks," "anti-quarks," "gluons," and all the rest, now become labeled with suspiciously subjective qualities such as "color," "flavor," "strangeness," and "charm." These are objects, as one contemporary physicist admits, that may never be directly observed; that could be compared to poetry, in that they need not "mean, but be"; about which one ought not ask "what" they are, but "when" they are.[9] Another suggests that quarks might better be regarded as events, rather than as particles.[10] Nearby in the subnuclear zoo we find the putative "tachyons"—particles that travel faster than light, have imaginary rest mass, lose energy

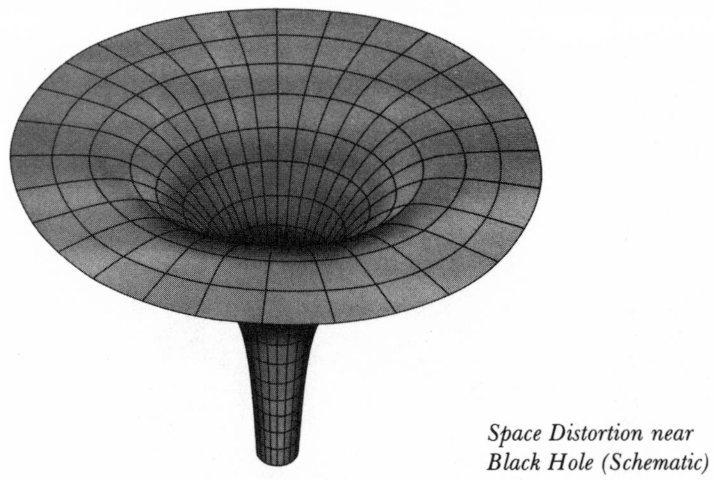

Space Distortion near Black Hole (Schematic)

when increasing in speed, and can reach infinite velocity, at which point they have zero energy but finite momentum. Then there are the even more controversial "anomalons"—nuclear fragments with anomalously strong interactive propensities that are difficult to replicate, since their characteristics seem to vary from one laboratory to another, and whose existence also raises some fundamental questions about basic statistical calculation.

In grasping for resolution of these enigmas, it is important to note that similarly ethereal physical objects and systems have been propounded far outside of the atomic and nuclear domains. For example, if we turn our eyes to the other extreme of reality's scale—to astrophysics, cosmology, and the dynamic universe—and doggedly follow our established scientific formalisms, we are inescapably confronted by another assortment of mind-stretching concepts. Perhaps most notable of these is the distortion of the metric space-time grid forecast by general relativity theory to derive from intense concentrations of substantive mass. This distortion reaches its climax in such curiously Freudian concepts as the "big bang" model of cosmic creation and the "black hole"—the most powerful of physical entities, devoid of all superficial detail, from whose influence no object or signal can escape, and for whose representation a new brand of "black-hole physics" must be posited.

May we not now reasonably ponder that if this powerful space- and time-bending property we call "mass" ultimately traces down to "particles" that we can only experience rather than observe, that we must treat as events rather than as substance, and that we can describe only in quasi-poetic, anthropomorphic terms, then it may indeed be consciousness that establishes physical reality, even on the cosmological scale? Are our minds being stretched, or are our minds doing the stretching? Perhaps another step needs to be taken along the road of scientific conceptualization that has brought us from causal particulate mechanics, to intangible wave/particle dualism, to probabilistic determinism, to observational realism. It is the step that explicitly acknowledges consciousness as constructor of the same reality it perceives, ponders, and postulates.

Consciousness (SHU) Distinguishing Earth (GEB) from Sky (NUT)

7

SCHOLARLY STRETCH

The overarching theme of the "Scholarly Stream" review of Section I-3 was the relatively recent segregation of science from philosophy, and more specifically of physics from metaphysics, following millennia of inseparable, symbiotic activity. The unfolding pattern of research evidence on consciousness-related anomalies of many forms suggests that this segregation has now become so extreme as to be counterproductive, posing a lame posture for further advance of either sector. There can be no quarrel that hard-nosed science of the sort envisaged by Francis Bacon, divorced from theological premise, political persuasion, or personal bias, and based solely on demonstrable empirical data and factual argument, has totally revolutionized and immensely benefited human existence over the past few centuries. It is equally evident that this progress has also entailed numerous environmental, social, and personal penalties, but these are not the primary basis of our concern. Rather, the point at issue is that modern science, by arbitrarily excluding all metaphysical parameters from its proper purview in the name of rigor and objectivity, is in danger of throwing out its intellectual and technological babies with the bathwater.

This shortfall was eloquently presaged by William James, nearly a century ago:

> The spirit and principles of science are mere affairs of method; there is nothing in them that need hinder science from dealing

successfully with a world in which personal forces are the starting point of new effects. The only form of thing that we directly encounter, the only experience that we concretely have is our own personal life. The only completed category of our thinking, our professors of philosophy tell us, is the category of personality, every other category being one of the abstract elements of that. And this systematic denial on science's part of personality as a condition of events, this rigorous belief that in its own essential and innermost nature our world is a strictly impersonal world, may, conceivably, as the whirligig of time goes round, prove to be the very defect that our descendants will be most surprised at in our boasted science, the omission that to their eyes will most tend to make *it* look perspectiveless and short.[11]

Similar exhortations have been voiced by many other visionary scholars. From the philosophical pen of Henri Bergson:

> Science and metaphysics therefore come together in intuition. A truly intuitive philosophy would realize the much-desired union of science and metaphysics. While it would make of metaphysics a positive science—that is, a progressive and indefinitely perfectible one—it would at the same time lead the positive sciences, properly so called, to become conscious of their true scope, often far greater than they imagine. It would put more science into metaphysics, and more metaphysics into science.[12]

From the astrophysical perspective of James Jeans:

> . . . the physical theory of relativity has now shown that electric and magnetic forces are not real at all; they are merely mental constructs of our own, resulting from our rather misguided efforts to understand the motions of the particles. It is the same with the Newtonian force of gravitation, and with energy, momentum and other concepts which were introduced to help us understand the activities of the world—all prove to be mere mental constructs, and do not even pass the test of objectivity.[13]

And from the theological insight of Teilhard de Chardin:

> The true physics is that which will, one day, achieve the inclusion of man in his wholeness in a coherent picture of the world.[14]

Thus, there is little new in our plea for some reintegration of the objective and subjective domains of experience and knowledge, except that the proposition is now supported by a substantial body of research evidence that demonstrates pragmatic effects of such synthesis, and suggests an approach to theoretical formulation for such rapprochement of scholarly mind and spirit.

"The marvelous complexity of forces which appear to control matter, if not actually to constitute it, are and must be mind-products."

ALFRED RUSSEL WALLACE

The possibility that consciousness, through intention, can marginally influence its physical reality to a degree dependent on its subjective resonance with the system or process in question, has implications that could extend well beyond those portions of engineering science, parapsychology, statistics, and quantum mechanics directly involved in this research, into the life sciences, the social sciences, and the humanities as well. For example, if we look past the technological provinces of microelectronics, artificial intelligence, and information processing, the next evident domain for further assessment of this influence would be that of the living organism. In principle, much of the experimental and theoretical methodology described in Sections II-IV could be directly applied to comparable studies of the interaction of human consciousness with simple biological systems and processes, particularly those appearing to rely on some form of probabilistic behavior, such as bacteria, algae, or sperm. While only a few basic experiments of this class have been performed and the data are yet too sparse to justify generic claims, evidence has indeed been accumulating to indicate that quite similar marginal biases of behavior can be effected in living systems as well.[15]

If consciousness effects in this domain can be more fully established, the implications and applications could range from molecular biology and genetic chemistry on one extreme, to general evolutionary theory on the other. The possibility that cell biology might entail a volitional or teleological component

Koko and her kitten

beyond random adaptation would constitute a major branch point in the comprehension and representation of the organizational capability of all classes of living systems and would inevitably force inclusion of motivational factors in their assessment. Again, this is far from a new proposition; aspects of it can be found in Lamarckian models of evolution, in theories of complex and dissipative structures,[16] in Bergson's concept of an *élan vital*, and in the controversial proposition of "morphogenetic fields."[17] But now we may be closer to illuminating such issues, and possibly even to resolving them, on firm research grounds.

A second important benefit of such research into biological anomalies could be in unifying some of the controversial but enduring propositions relating to nontraditional medical practice and health care. Many of the claims for spiritual healing and diagnosis, holistic health strategies, placebo effects, hypnotic techniques, pet-therapy, plant care, and various psychotherapeutic procedures seem to entail the same two common denominators we find in the physical system anomalies, namely the roles of resonance, or empathy, and volition, or will. Even in more traditional medical practice it is becoming increasingly evident that such nonphysical parameters can play important roles in immunology and healing. If the statistical behavior of microscopic physiological systems and processes can be shown to be correlated with the intentions of human participants, the base may have been laid for rational explication, and possibly for systematic enhancement, of a variety of currently unorthodox practices.

The fields of cognitive and behavioral psychology, through their alliances with neurophysiology and biochemistry, may also share in these biological implications. Like the physical sciences, academic psychology has tended increasingly to segregate itself from the subjective and metaphysical dimensions of its subject matter, and even from its own clinical practice, preferring to concentrate on the more tangible and quantitative aspects of perception, memory, and observable behavior. The seminal works of such pioneers as James, Freud, and Jung are largely overlooked in its modern curricula, and little effort is made to include the psyche, or spirit—or even consciousness itself—in the prevailing models of human behavior. If firm evidence for the influence of consciousness on either living or nonliving systems can be sustained, and if the resonant bond model continues to be supported by experimental data, then the purview of modern psychology, like that of physics, may require extension to accommodate some form of metaphysical mechanics.

"The relationship between doctor and patient, especially when a transference on the part of the patient occurs, or a more or less unconscious identification of doctor and patient, can lead to parapsychological phenomena. I have frequently run into this."

CARL JUNG

This dimension seems better recognized by many clinical practitioners who have acknowledged the role of love or empathy in effective therapy. In our numerous discussions with psychiatrists and therapists, instances of apparent psychokinetic, precognitive, or clairvoyant anomalies encountered in their practices have repeatedly been cited. Practitioners of athletic psychology have also confirmed popular impressions that the most gifted performers innately possess or develop extraordinary abilities to foresee the courses of action, to protract time scales, or to achieve transcendent resonances with their teammates or equipment. Many of the "new-age" counseling, therapy, and actualization programs rely heavily on empathy and volition in the reformulation of long-standing habits of negative behavior. The challenge is to sift, meld, and deploy all of this lore into a more systematic and analytical reservoir of scholarly comprehension, bounded on all sides by rigorous research results.

The implications of such metaphysical components may also extend to other social sciences. In the fields of sociology, anthropology, politics, and economics, such intangible factors as social belief and value systems, charismatic leadership, group management, public opinion, and other aspects of the social

Antony's Funeral Oration

ambience, or zeitgeist, have long been appreciated, but seldom formulated systematically. Although the processes here may be more complex, the roles of human volition and resonance at societal levels might now be more consistently addressed in the light of quantitative research data and wave-mechanical models of collective consciousness. For example, deliberate investigations of group interactions with physical systems and processes, beginning with multiple operator experiments like those described in Section II-6, or others testing various implications of the model for group behavior, could be instructive.

Moving yet further from the technical base into the areas of the cultural humanities, a number of diverging streams can be identified that may be influenced, and possibly even reintegrated, by acknowledgment of the research data. First, it should be noted that classical art, music, literature, and drama have never been bashful in including metaphysical aspects in their respective creative theses. Mythical and spiritual allusions integral to their themes and characterizations can be found in every age and genre of artistic expression, and much of this material falls well within the phenomenological catalogue addressed in the scientific sector. From this perspective, the research results can only serve to strengthen such creative license.

Beyond this, it should be noted that modern technology has recently exerted at least two important influences on the creative humanities. On the one hand, through mass media, higher quality reproductions, and enhanced programs of education, many forms of artistic expression have become more accessible to broader segments of the population, and their influence has been correspondingly increased. On the other hand, this same technology has impelled the scholarly activities in these areas progressively more toward critical dissection, compartmentalization, and artificial reproduction, and away from intuitive, inspirational, and holistic composition.

Perhaps related to this influence has been the role of the contemporary creative arts as vehicles of protest against the prevailing materialistic social ethics, vis-à-vis the primal per-

Le Modèle Rouge *(Magritte)*

sonal needs for aesthetic inner experience. Twentieth-century art, literature, and music are rife with distortions, dissonances, surrealities, and contradictions that could be regarded as anomalies in their own genres of expression, and may well be indicative of a fundamental inability of the modern paradigm to accommodate artistic creation in an excessively reductionistic and mechanistic world. Stated in terms of the consciousness correspondence principle, there may be insufficient room in this short-wavelength society for the longer-wavelength components of the human psyche to express themselves without distortion. To this intuitive sense of imbalance, the research adds the injunction of the consciousness uncertainty principle, arguing for better complementarity of the analytical and aesthetic conjugates of life.

The most obvious and explicit incorporation of metaphysical phenomena within an established scholarly domain clearly prevails in the disciplines of academic theology. Virtually all contemporary religions, like most of their predecessors, trade heavily in such mystical attitudes and activities as faith, hope, love, prayer, and sacrifice. Research results like those offered here may add another perspective to such elusive metaphysical concepts. Indeed, the relevance and vitality of theology in an increasingly analytical and technological age might well benefit from its own appropriate forms of scientific research, in much the same way that modern diagnostic and computational techniques have advanced factual comprehension and even aesthetic appreciation in such traditionally impressionistic fields as art, music, and literature. In particular, some broadening of the purview of theological research from its traditional literary and archaeological arenas, to include controlled scientific study of the roles of human attitudes, strategies, and beliefs in the establishment of reality, could provide a fuller appreciation of the dimensions, mechanics, and purposes of the sublime processes.

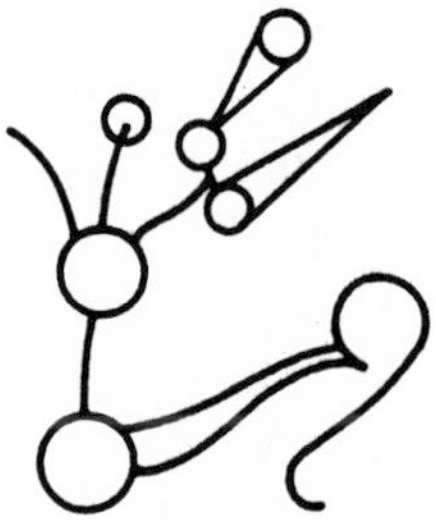

The Philosopher's Stone

8

MODERN MYSTICAL MAN

The memorial service for J. B. Rhine, held in the Duke University Chapel a few years ago, was attended by an assortment of family members and friends, former and present students and staff from his laboratory, professional colleagues and admirers from the parapsychological community, a few appropriate dignitaries, some curious campus residents, and a number of faculty and administrators who, while respecting the man's personal characteristics and professional integrity, had never been fully persuaded by his research. Following a succession of eulogies and vignettes outlining various aspects of Rhine's life and work, the final participant stepped to the pulpit with an unexpected challenge. "Each of us," he said, "will now have the opportunity to decide whether we really believe in the message of J. B. Rhine.—Let us pray."

Stripped of its particular personal and parochial context, his point was fundamental and telling. One cannot sincerely pray and still logically deny all possibility of influence of human volition on physical events. Yet modern life has become so compartmentalized that our intellectual and spiritual activities tend to be carried on in separate contexts, with very little relevance, let alone reinforcement between them. We attend our lodges, encounter groups, holistic health, martial arts, and religious group meetings on the weekends or in the evenings, and the following morning put on our working clothes to return to strictly analytical, increasingly digital, virtually impersonal manipulation of our business or professional responsibilities. And all the while, well beneath our starched collars or laboratory coats, frail and timid vestiges of our spiritual selves continue to dream, to hope, to love, and to pray.

At the risk of sounding like *Star Trek*'s Mr. Spock, is it logical for humans to indulge in wishing, or praying, or loving, yet to doubt the efficacy of such activities? Or if their empirical effectiveness is acknowledged, why should such phenomena be exempt from scientific study, or from systematic application in our pragmatic or scholarly areas of activity? Why cannot our mystical and mental selves coexist in both practical and spiritual affairs; indeed, why can they not simply coalesce?

Again, the plea for a mutually enhancing union of these two unique human capacities is not new. The greatest minds of many disciplines and many eras have voiced the need repeatedly, none more incisively or beautifully than Albert Einstein:

> Now, even though the realms of religion and science in themselves are clearly marked off from each other, nevertheless there exist between the two strong reciprocal relationships and dependencies. Though religion may be that which determines the goal, it has, nevertheless, learned from science, in the broadest sense, what means will contribute to the attainment of the goals it has set up. But science can only be created by those who are thoroughly imbued with the aspiration towards truth and understanding. This source of feeling, however, springs from the sphere of religion. . . . The situation may be expressed by an image: Science without religion is lame, religion without science is blind.[18]

Earlier, in the context of the quantum mechanical metaphor, we spoke of the profound concept of complementarity—that ecumenical scientific perspective needed for philosophical resolution of the wave/particle dilemma, the uncertainty principle, and the role of the observer—and we discussed its extension into broader philosophical domains. In the interplay of spirit and intellect, we deal with the ultimate conjugate perspectives whereby consciousness triangulates all of its abstract experience, and these can be deployed either in awkward, encumbering contradiction or in mutually fulfilling complementarity. The desirability of the latter approach has, of course, long been propounded by pen and pulpit, and in many quarters accepted on faith in abstract terms. The message of this book is simply to confirm the efficacy of human volition, deployed in self-effacing resonance with a given system, process, or another consciousness, in specific, pragmatic, and quantitative terms. The relevance of such confirmation to individual spiritual belief and ethic is, of course, a highly personal matter.

"A religion contradicting science and a science contradicting religion are equally false."

P. D. OUSPENSKY

Thus, as we attempt to assess the relevance of this work to the vector we called "Man and the Mystical," we reach much the same realization as for the "Scholarly Stream"; namely, each of these should be deflected toward reentwinement with the other, and their future courses should reflect this complementarity. This synthesis need not be forced or difficult, for the two perspectives share much common heritage, purpose, and strategy; it is mainly the conceptual vocabularies that differ. Science, viewed from the mystical perspective, is the cosmic questioning force seeking comprehension of itself through systematic generation, processing, and interpretation of information. Through successively finer discriminations and correlations of its experience, it strives objectively to understand the nature of reality. The mystical, viewed from the scientific perspective, is a yearning to transcend the constraints of the physical world by seeking direct experience of the cosmic mysteries through subjective integration of the self into their milieu.

Both scientific and mystical knowledge are manifested by their consequences: science via engineering, technology, and medicine; mystical insight via human values, perspectives, and behavior. Whether we deal with objective study of subjective phenomena, or vice versa, our experiential eigenfunctions reflect both their environmental profiles and their own intrinsic wave properties. If these components are cast in separated, provincial coordinates, their interaction will be confused and sterile. If they are established in unified, holistic form, the resulting experience and understanding will be balanced and fertile. No one recognized this benefit better than the first of the modern physicists, Max Planck:

> In truth, science and religion present no contrasts but rather for every person of serious bent they need each other for mutual supplementation. It is surely no accident that precisely the greatest thinkers of all times were also deeply religious. . . . Only out of the concurrence of the powers of the mind and of the will has there evolved philosophy's ripest, most precious fruit: ethics.[19]

9

ORGANIZATION AND ORDER

The concepts, vocabularies, and heritages of the scholarly vectors just reviewed are sufficiently disparate to preclude any more comprehensive summary of their possible future deflections, but one common theme pervades their various dimensions. It is the role of consciousness in organizing and ordering information or, conversely, in reversing random tendencies toward disorganization, disorder, and chaos. This process may be further illustrated in terms of a few additional physical metaphors drawn from the areas of statistical thermodynamics and information theory.

> **"To have order, you must have randomness, because where there is no randomness order cannot manifest itself."**
>
> ALAN WATTS

One of the most broadly useful concepts in these fields is a property called entropy, which is an index of the degree of disorder of any proliferate system, inversely related to the amount of information available about it. As a crude example, imagine a rectangular box, inside of which, at the left end, are placed a number of white marbles and, at the right end, an equal number of black marbles. If, without looking, we reach into the left end to retrieve a marble, we should, with a high degree of certainty, find a white one; at the right end, we should find a black one. This state of segregation of the marbles is one of low entropy, or, equivalently, of high information.

If, however, the box is tumbled for a while, the marbles will mix, and any later attempt to retrieve one of a given color, at either end, can be predicted with considerably less certainty. Eventually, after thorough mixing, the probability will approach .50/.50. This is the state of maximum randomness or entropy or, alternatively, of lowest information or order. Note that this mixing process is irreversible, in the sense that no further mixing is likely to restore the original segregated arrangement, or indeed even to improve the .50/.50 probability in any predictable fashion. Only by external intervention, such as manual rearrangement of the balls, can this now be achieved.

The process just described is encompassed by a general physical principle—usually termed the Second Law of Ther-

"All the king's horses and all the king's men
Couldn't put Humpty together again."

modynamics—that states that any isolated system will spontaneously decay, in some statistical fashion, toward its highest entropy, lowest order configuration. This principle pertains most directly to all physical systems, whether classical or quantal in nature, but has been extended, at least metaphorically, to biological, sociological, and psychological situations as well. In all applications, it is essential to specify and to maintain the particular isolated system of interest, since the spontaneous decay of order can always be countermanded by external intervention.

The results of our laboratory experiments could be construed to violate this principle, in the sense that certain maximum entropy or minimum information distributions of bits have been altered to slightly more ordered, lower entropy distributions, solely in response to human intention. In the man/machine experiments, for example, the PK^+ and PK^- output distributions achieved under corresponding operator intentions are slightly more ordered in their correspondence to the +, −, +, − . . . reference sequence than are those obtained in calibrations. Similarly, in the PRP studies, a state of zero information about the target has been marginally improved, solely by percipient intention to perceive it. In neither case has any known external physical mechanism influenced the system.

> **"Mind filters out matter from the meaningless jumble of qualities, as the prism filters out the colours of the rainbow from the chaotic pulsations of white light."**
> ARTHUR EDDINGTON

To resolve this contradiction, one of three concessions must be made: a) the Second Law is not universally valid, that is, these particular isolated systems possess a self-organizing capability; b) these systems are not totally isolated, that is, some influence has entered or left them; or c) these systems should be redefined to include the operator or percipient as integral parts of them. Option a) clearly requires further specification of the self-organizing mechanism of these special systems. Option c), which is a special case of b), leaves unresolved the nature of the interaction between the operator or percipient and the machine or target.

Many other examples of apparently entropy-defying systems or processes could be cited. Biology provides abundant cases of organisms that spontaneously progress from lower to higher states of organization without evident external impetus. Certain cosmological models, especially those addressing the creation of the universe and of life within it, could be construed to leave gaps in the applicability of the entropy principle. While some of these apparent exceptions can be circumvented by elaborate theoretical argumentation, more parsimonious explication could follow from postulation of an entropy-reducing influence somehow associated with individual or collective consciousness. Cast in sufficiently generic form, such a postulate could accommodate not only physical and biological anomalies of this type but, by metaphor, a broad range of psychological, philosophical, social, and religious effects as well. Indeed, it could acknowledge consciousness, in the most general sense, as a universal organizing factor in the cosmos.

"For if physical events are indeterminate and the future is unpredictable, then perhaps the unknown quantity called 'mind' may yet guide man's destiny among the infinite uncertainties of a capricious universe."

LINCOLN BARNETT

Again we would submit that this apparently radical proposition may entail little more than a specialized extension of a common cognitive process. The basic business of consciousness is to select and order information from the chaos of stimuli available to it, in order to execute some utilitarian or satisfying function. Whether assembling rudimentary pencil marks or vocal elements into letters, words, and sentences; visually or auditorially transcribing those combinations into meaningful concepts; or extracting coherent patterns of information and experience from any other surfeit of physical or emotional stimuli, consciousness is a master of intellectual organization and order. Indeed, the entire psycho-

The Creation of Onogoro from the Ocean of Chaos (19th Century)

logical process of learning, along with all formal and informal educational systems that support it, could be regarded as a grand entropy-reducing exercise. In this light, the ability to rearrange a few ones and zeros within a random binary string, or to compose useful information about a remote target scene, perhaps becomes less extraordinary, especially when encouraged by a favorable cultural ethic or personal paradigm.

Any name that might be proposed for this organizing capacity of consciousness would inevitably be confounded by alternative connotations. We might, with de Broglie, call it

"Love, in a very general sense"; with Bergson, "life force"; with the theologians, "Holy Spirit" or "Divine Mind"; or with the mystics, "Unity" or "Self," but by any title it encompasses the potential that Teilhard de Chardin envisaged when he wrote:

> Someday, after we have mastered the winds, the waves, the tides and gravity, we shall harness for God the energies of love. Then for the second time in the history of the world, man will have discovered fire.[20]

Whatever may be assigned to this chaos-reducing capability of consciousness, if L is its symbol, the process it defines is clearly of the form

$$L = -\Delta s,$$

where $-\Delta s$ denotes the decrease in entropy of the relevant system, the increase in its information, the establishment of its reality.

10

PERHAPS PRINCIPLES

The tale is told of the king who charged his wisest court scholars to prepare the ultimate statement on the nature of reality. After many years of heroic effort, these savants provided an immense intellectual tome, but were rebuffed with an instruction to digest it all into just one paragraph. Severely daunted, they nonetheless returned to their labor and after several more years of excruciating debate brought forth their consensus condensation. "Still too long," grumbled the king. "Give me just *one word.*" In great weariness and frustration, the group again resumed their deliberations, and it was not until the close of their lives that the few surviving scholars shuffled forward to present, with trembling hands, a battered, heavily erased parchment. On it was indeed inscribed a single word:

Perhaps.

This, too, has been a book about "perhaps." Not the common passive "perhaps" of equivocation, indecision, or ignorance, but a more assertive "perhaps" affirming personal opportunity for participation in the daily determinations of the cosmos. The description of that opportunity, along with some hints for its activation, has been based not on raw metaphysical speculation or unsubstantiated precedent, but on a standard scientific two-step of physical experimentation and metaphoric transposition of established theoretical concepts. And the "perhaps"

Sri Yantra

thus defined—the margins of reality to which our consciousness may lay claim—has turned out to be no more than a natural consequence of the manner in which it assembles its information and poses its questions.

Two of the most effective tactics consciousness employs in assimilating and utilizing its incessant influx of environmental stimuli are the processes of discrimination and association. In endeavoring to reduce the entropy of our experience we discriminate between "apples" and "oranges" on one set of criteria, and invoke their common features on another, in an iterative progression of logic that converges toward specification of the items of interest. In our anomalies experiments, for example, the distinctions between individual operator achievements, between PK^+ and PK^- performance, between instructed and volitional or automatic and manual conditions, or among the other technical parameters are critical in cataloguing the data; yet, the observed similarities of repeated efforts by a given operator, or of his signature on a variety of devices, or of the statistical character of the REG, RMC, and PRP results, are equally essential to comprehension of the process. At a more rudimentary level, the distinction between the binary digits 0 and 1, or between the + and − samples, is central to all of the data generation and processing, but the similarity of their output distributions is the basis for all subsequent interpretation. On the theoretical side, as well, the differentiation of the quantized states of physical atomic structures is fundamental to their comprehension and representation, but it is the pattern of similarities between these and the functioning of human consciousness that underlies our metaphorical model.

In any domain of contemplation, consciousness tends to weave structures of logic by successive application of such discrimination and association—first separating, then correlating; differentiating, then integrating; analyzing, then synthesizing—until a serviceable net of orthogonal conceptual fibers has been woven wherein it may collect, sift, and sort the plethora of potential information available from its environment. And since these strands of discrimination and association are themselves devised by the consciousness, they must be complementary, in Bohr's fullest sense of the term, and therefore constrained at every junction of their net by the uncertainty of mutual specification Heisenberg identified. From that array of uncertainty, distributed throughout the consciousness web, devolve the margins of reality we encounter in our research, in our model, and in our daily lives.

The most striking examples of this flexible discrimination/association dynamic of consciousness are the resonant bonds it establishes with people, places, things, and processes. To achieve such bonds, whether in physical or consciousness space, it is first necessary to acknowledge that there are distinct partners. That distinction established, however, the individual identities must then be at least partially surrendered to the bonded state if the exchange energy is to be activated. Thus, successful strategy for anomalies experimentation involves some blurring of identities between operator and machine, or between percipient and agent. And, of course, this is also the recipe for any form of love: the surrender of self-centered interests of the partners in favor of those of the pair.

"Love is a kind of exalted but unspecialised telepathy;—the simplest and most universal expression of that mutual gravitation or kinship of spirits which is the foundation of the telepathic law."

F. W. H. MYERS

In this sense, once again, the requisite approach to either commitment must be balanced between the active and the passive, the giving and the receiving, the doing and the being. To the rhetorical questions posed by our colleagues regarding

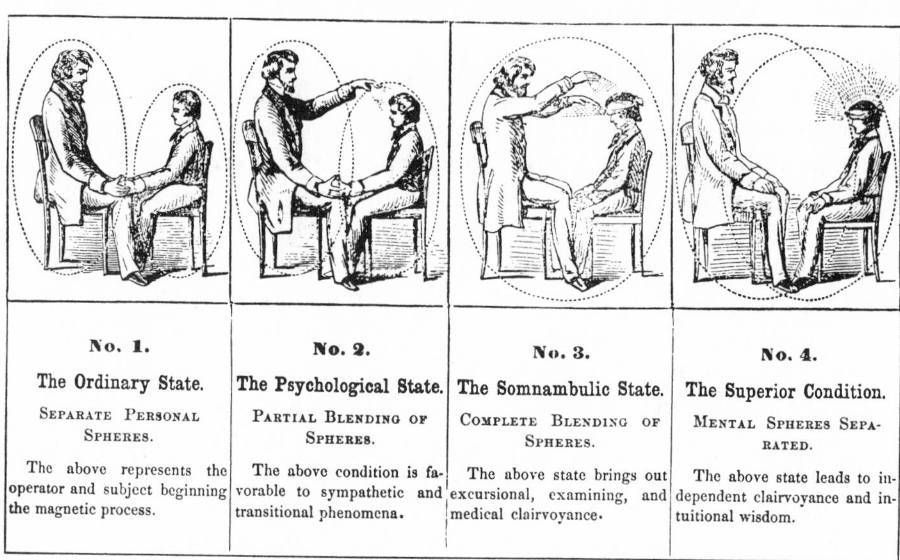

No. 1.
The Ordinary State.
Separate Personal Spheres.
The above represents the operator and subject beginning the magnetic process.
No. 2.
The Psychological State.
Partial Blending of Spheres.
The above condition is favorable to sympathetic and transitional phenomena.
No. 3.
The Somnambulic State.
Complete Blending of Spheres.
The above state brings out excursional, examining, and medical clairvoyance.
No. 4.
The Superior Condition.
Mental Spheres Separated.
The above state leads to independent clairvoyance and intuitional wisdom.

"Four Steps of the Magnetic Blending of the Minds" (19th Century)

our motivation in pursuing this research and the implications of its results, we could have added a third equally frequent and equally difficult query—something like: "But just *how* do your operators and percipients achieve their anomalous results?" or simply, "What is it that they *do*?" By their format, such questions already miss the major point. It is not only what one *does* that is critical, but also what one *is*. It is not just the operator's tactics, but his state that prevails. Results follow naturally from immersion in the task, as much as from attention to it. Such phenomena are not deterministic in the classical sense; they are created in the delicate dialogue between objective discrimination and subjective association.

This book itself has deployed a similarly woven structure of discrimination and association to assemble its case. The scholarly vectors that motivated and defined the research program and subsequently provided the basis for its assessment comprise a differential set of topical spokes around which have been woven unifying spiral strands of common features, metaphors, and other forms of association to build a conceptual basket for consideration of the role of consciousness in physical reality.

It is our suggestion that this particular structure has universal implications for all consciousness processes, for in its assembly of rules for the *study* of reality, it automatically establishes the template for the *experience* of reality.

Because these rules of reality are equally pertinent to any of the topical spokes that radiate through the epistemological pattern, they can equally well be summarized in the language of any one of them, and then extended by metaphor to all the others. Since the quantum mechanical nomenclature has served us well throughout, let us also propose our rule summary in this vocabulary. In fact, we need little more for this purpose than a brisk restatement of the generalized quantum mechanical principles that are, in fact, the basic guidelines for discrimination and association in their own, and thus in any other, conceptual genre:

Reality Rules

Complementarity

Consciousness possesses both particulate and wave-like characteristics, from which devolve an assortment of many other pairs of conjugate perspectives, such as doing/being, observation/participation, objective/subjective, or function/structure. Such pairs are not polar opposites, nor are they mutually contradictory; they are alternative consciousness states that complement one another and are bound by the uncertainty and correspondence principles.

Uncertainty

Simultaneous specification of complementary pairs of consciousness properties is mutually constrained; precise identification of one inevitably obscures the other. Optimization of creativity, analytical skill, artistic performance, or anomalous influence requires judicious trade-off between the complements appropriate to the task. For meticulous analytical achievement, emphasis may need to be toward the reductionistic, objective, "short-wavelength" characteristics; for aesthetic, inspirational, transcendent experience, the holistic, subjective, "long-wavelength" processes may need to dominate. But simultaneous maximization of both components is not an option; in precluding such bilateral precision, the uncertainty rule defines the margins of reality.

Indistinguishability

Sharing or blurring of personal identity in interactions with persons, devices, or other relevant portions of the environment can lead to resonant-system behavior sufficiently different from the individual expectations to be regarded as "anomalous." This behavior is, however, perfectly normal for the bonded systems; it is only anomalous in reference to a linear combination of the normal behaviors of the separate components.

Exclusion

The strongest bonds of consciousness are those in which the receptive/assertive roles are fulfilled by the participants, in a dynamic give-and-take. If both participants pursue the same role, an anti-bond or aversion may arise. In either case, anomalous effects may appear, but of different character.

Correspondence

The wave-like aspects of consciousness, and all of the anomalies associated with them, only manifest when its wavelength, as defined by the rate of relevant information processing, is comparable with or longer than the dimensions of its environmental context. Whenever this wavelength is much shorter than those environmental dimensions, behavior and experience revert to common particulate forms. Thus, the preponderance of potential anomalous effects is suppressed by the common analytical traffic of the mind.

Again, these basic guidelines for the composition of reality could be, and indeed have been, articulated in many other languages and contexts, but in whatever form they may be cast, they define the principles of the magnificent "perhaps" the cosmos affords to all who have the desire and courage to play with it. Collectively they are tantamount to an inextricable association of the processes of the physical world with the processes of the mind. As Arthur Eddington concluded, toward the close of his own scientific journey:

> Not once in the dim past, but continuously by conscious mind is the miracle of the Creation wrought.[21]
>
> All through the physical world runs that unknown content, which must surely be the stuff of our consciousness. . . . Where science has progressed the farthest, the mind has but regained from nature that which the mind has put into nature.
>
> We have found a strange foot-print on the shores of the unknown. We have devised profound theories, one after another, to account for its origin. At last, we have succeeded in reconstructing the creature that made the foot-print. And Lo! it is our own.[22]

To which the research described in this book can only add, humbly but affirmatively,

Perhaps.

Appendix, Notes, Sources for Sidebar Quotations, Illustration Credits, Suggestions for Further Reading, Acknowledgments, Index

Eye Goddess (Syria, ca. 2800 BC)

APPENDIX

TABLE A

REG Data Summary by Operator

		PK⁺						PK⁻					
Opr.	No. of Series	No. of Trials	Mean	*z*-Score	Prob.	No. of Series $p < .05$*	No. of Series $p < .5$	No. of Trials	Mean	*z*-Score	Prob.	No. of Series $p < .05$*	No. of Series $p < .5$
10	15	55,100	100.082	2.729	.003	3	12	55,050	99.896	−3.459	3×10^{-4}	4 (1)	13
14	3	8,000	100.070	0.885	.188	1	2	7,800	99.872	−1.603	.054	1	3
16	3	7,500	100.070	0.856	.196	–	2	7,500	99.763	−2.903	.002	1	3
19	1	2,950	100.030	0.232	.408	–	1	2,800	100.042	0.313	(.377)	–	–
20	3	7,550	100.087	1.064	.144	–	2	7,450	99.979	−0.262	.397	–	2
21	1	2,700	100.044	0.321	.374	–	1	2,300	100.156	1.056	(.146)	–	–
29	1	2,500	99.912	−0.625	(.266)	–	–	2,500	100.046	0.322	(.374)	–	–
30	2	5,000	100.026	0.262	.397	–	1	5,000	99.939	−0.606	.272	–	2
33	1	4,000	99.868	−1.178	(.119)	–	–	2,500	99.928	−0.512	.304	–	1
36	2	5,000	99.978	−0.218	(.414)	–	1	5,250	100.068	0.695	(.244)	–	1
41	5	13,450	100.023	0.373	.355	–	3	15,050	99.984	−0.273	.392	– (1)	4
42	1	2,700	100.094	0.691	.245	–	1	2,300	100.031	0.212	(.416)	–	–
44	2	5,300	99.781	−2.255	(.012)	– (1)	–	6,200	99.918	−0.914	.180	–	2
49	2	4,950	99.871	−1.284	(.099)	–	–	5,050	100.072	0.722	(.235)	–	1
53	3	7,550	99.937	−0.778	(.218)	– (1)	1	7,450	99.981	−0.236	.407	–	2
55	13	43,300	100.018	0.544	.293	2 (1)	7	43,400	100.028	0.818	(.207)	– (1)	5
59	2	5,100	100.046	0.461	.322	–	1	3,900	99.923	−0.684	.247	–	2
64	2	5,500	99.940	−0.625	(.266)	–	1	4,500	100.042	0.403	(.344)	–	1
65	1	2,600	100.207	1.489	.068	–	1	2,400	99.956	−0.303	.381	–	1
66	2	7,950	100.003	0.041	.484	–	1	7,050	99.930	−0.830	.203	–	1
68	1	4,650	99.955	−0.429	(.334)	–	–	5,350	100.005	0.048	(.481)	–	–
70	3	7,700	99.963	−0.459	(.323)	–	1	7,300	99.922	−0.940	.174	–	2
80	2	7,500	100.185	2.272	.012	1	1	7,500	99.871	−1.581	.057	1	2
82	2	4,100	100.193	1.745	.041	1	2	5,250	100.060	0.615	(.269)	–	–
84	1	2,500	100.102	0.724	.235	–	1	2,500	99.765	−1.663	.048	1	1
85	1	5,100	100.088	0.891	.186	–	1	5,300	100.180	1.851	(.032)	– (1)	1
88	1	2,750	100.169	1.254	.105	–	1	3,000	100.084	0.648	(.258)	–	–
90	1	2,500	100.088	0.619	.268	–	1	2,500	99.968	−0.229	.409	–	1
92	2	7,400	99.941	−0.722	(.235)	1 (1)	1	5,400	100.087	0.903	(.183)	– (1)	1
93	2	5,000	100.144	1.440	.075	–	2	5,000	100.019	0.194	(.423)	–	1
94	4	10,000	100.056	0.796	.213	1	2	10,000	99.939	−0.863	.194	–	3
96	1	2,250	100.073	0.492	.311	–	1	2,750	100.085	0.634	(.263)	–	–
97	1	2,500	100.110	0.781	.218	–	1	2,500	99.935	−0.461	.322	–	1
All	87	262,650	100.037	2.666	.004	10 (4)	53	259,800	99.966	−2.444	.007	8 (5)	56

REG Data Summary by Operator—(Continued)

	Baseline							ΔPK				
Opr.	No. of Series	No. of Trials	Mean	z-Score	Prob.	No. of Series $p < .05$	No. of Series Mean > 100	No. of Trials	z-Score	Prob.	No. of Series $p < .05$*	No. of Series $p < .5$
10	13	59,100	100.011	0.389	.349	–	8	110,150	4.375	6×10^{-6}	4	14
14	3	7,250	99.936	−0.774	.219	–	1	15,800	1.756	.040	–	3
16	3	7,500	100.024	0.292	.385	–	1	15,000	2.658	.004	1	3
19	1	2,500	100.044	0.311	.378	–	1	5,750	−0.052	(.479)	–	–
20	3	7,500	99.956	−0.537	.296	–	0	15,000	0.940	.174	–	3
21	1	2,500	100.032	0.229	.409	–	1	5,000	−0.480	(.316)	–	–
29	1	2,500	100.076	0.540	.295	–	1	5,000	−0.670	(.251)	–	–
30	1	2,650	99.978	−0.162	.436	–	0	10,000	0.614	.270	–	1
33	1	2,500	100.053	0.376	.353	–	1	6,500	−0.607	(.272)	–	–
36	2	4,500	99.928	−0.679	.249	–	1	10,250	−0.650	(.258)	–	–
41	3	9,800	100.012	0.166	.434	–	1	28,500	0.455	.325	–	4
42	1	2,500	99.964	−0.255	.400	–	0	5,000	0.364	.358	–	1
44	2	3,000	99.982	−0.137	.446	–	1	11,500	−0.860	(.195)	–	–
49	2	5,000	100.078	0.784	.217	–	2	10,000	−1.417	(.078)	– (1)	–
53	3	7,500	99.961	−0.477	.317	–	2	15,000	−0.386	(.350)	– (1)	2
55	10	40,000	100.004	0.113	.455	–	6	86,700	−0.194	(.423)	1 (1)	6
59	1	1,100	100.119	0.559	.288	–	1	9,000	0.798	.213	–	2
64	1	5,500	99.908	−0.965	.167	–	0	10,000	−0.734	(.232)	–	–
65	1	2,500	100.172	1.213	.112	–	1	5,000	1.284	.100	–	1
66	2	7,500	99.964	−0.444	.328	–	1	15,000	0.599	.275	–	1
68	1	4,950	100.158	1.572	.058	–	1	10,000	−0.328	(.371)	–	–
70	3	7,500	100.081	0.998	.159	–	2	15,000	0.327	.372	–	2
80	2	5,500	100.160	1.676	.047	–	2	15,000	2.724	.003	1	1
82	1	5,000	100.014	0.144	.443	–	1	9,350	0.695	.244	–	2
84	1	2,500	100.092	0.653	.257	–	1	5,000	1.688	.046	1	1
85	1	5,550	100.107	1.129	.129	–	1	10,400	−0.698	(.243)	–	–
88	1	2,900	99.915	−0.649	.258	–	0	5,750	0.399	.345	–	1
90	1	2,500	99.820	−1.270	.102	–	0	5,000	0.600	.274	–	1
92	2	4,950	99.900	−0.997	.159	–	0	12,800	−1.135	(.128)	1 (1)	1
93	2	5,000	99.827	−1.730	.042	–	0	10,000	0.881	.189	–	2
94	4	9,500	99.961	−0.543	.294	–	2	20,000	1.173	.120	1	3
96	1	2,500	99.914	−0.611	.271	–	0	5,000	−0.140	(.444)	–	–
97	1	2,500	100.104	0.735	.231	–	1	5,000	0.878	.190	–	1
All	76	243,750	100.004	0.282	.389	–	41	522,450	3.614	2×10^{-4}	10 (4)	56

* Numbers in ( ) indicate results opposite to intention.

TABLE B

RMC Data Summary by Operator

				PK⁺ (right)					
Opr.	No. of Series	No. of Runs	BL Mean	Mean	S.D.*	*t*-Score*	Prob.	No. of Series $p < .05$	No. of Series $p < .5$
10	17	270	10.0328	10.0297	.0508	−0.987	(.162)	– (2)	6
14	1	20	10.0336	10.0284	.0497	−0.472	(.321)	–	–
16	3	30	10.0111	10.0246	.0496	1.497	.073	–	3
20	2	20	10.0002	10.0226	.0505	1.975	.031	1	2
41	7	90	10.0205	10.0240	.0533	0.620	.268	–	4
42	3	30	9.9913	10.0181	.0452	3.248	.001	1	3
44	1	20	10.0419	10.0361	.0457	−0.571	(.287)	–	–
49	1	10	10.0219	10.0291	.0462	0.488	.319	–	1
51	1	10	10.0047	10.0066	.0456	0.133	.448	–	1
53	1	10	10.0283	10.0214	.0682	−0.321	(.378)	–	–
55	20	300	10.0272	10.0283	.0497	0.374	.354	2 (1)	11
63	1	7	10.0207	10.0342	.0458	0.782	.232	–	1
64	1	10	10.0164	10.0219	.0476	0.364	.362	–	1
66	1	10	10.0210	9.9907	.0306	−3.142	(.006)	– (1)	–
68	2	40	10.0224	10.0180	.0453	−0.608	(.273)	–	1
69	1	11	10.0463	10.0253	.0679	−1.023	(.165)	–	–
70	4	40	10.0166	10.0278	.0541	1.310	.099	–	4
79	1	9	10.0214	10.0448	.0649	1.079	.156	–	1
84	1	10	10.0237	10.0348	.0376	0.937	.187	–	1
91	1	16	10.0290	10.0325	.0462	0.306	.382	–	1
93	3	31	10.0142	10.0219	.0492	0.875	.194	–	2
94	3	30	10.0041	9.9977	.0557	−0.636	(.265)	– (1)	2
All	76	1024	10.0245	10.0260	.0506	0.978	.164	4 (5)	45

RMC Data Summary by Operator—(Continued)

	PK^- (left)						ΔPK (right–left)**					
Opr.	Mean	S.D.*	t-Score*	Prob.	No. of Series $p < .05$	No. of Series $p < .5$	No. of Pairs of Runs	S.D.*	t-Score*	Prob.	No. of Series $p < .05$	No. of Series $p < .5$
10	10.0195	.0536	−4.076	3×10^{-5}	6	14	270	.0521	3.235	7×10^{-4}	7 (1)	11
14	10.0259	.0418	−0.822	.211	–	1	20	.0402	0.270	.395	–	1
16	9.9994	.0507	−1.256	.110	1	2	30	.0515	2.683	.006	1	3
20	10.0259	.0448	2.566	(.009)	– (1)	–	20	.0328	−0.461	(.325)	– (1)	1
41	10.0241	.0488	0.697	(.244)	–	3	90	.0530	−0.018	(.493)	– (1)	3
42	9.9990	.0502	0.838	(.204)	–	–	30	.0432	2.427	.011	1	3
44	10.0342	.0484	−0.713	.242	–	1	20	.0531	0.158	.438	–	1
49	10.0212	.0369	−0.058	.478	–	1	10	.0375	0.659	.263	–	1
51	10.0002	.0322	−0.436	.337	–	1	10	.0307	0.655	.264	–	1
53	10.0195	.0632	−0.444	.334	–	1	10	.0288	0.214	.418	–	1
55	10.0244	.0494	−0.985	.163	1	13	300	.0490	1.373	.085	2	13
63	10.0126	.0391	−0.550	.301	–	1	7	.0358	1.600	.080	–	1
64	10.0241	.0506	0.482	(.321)	–	–	10	.0523	−0.134	(.448)	–	–
66	10.0017	.0590	−1.038	.163	–	1	10	.0478	−0.728	(.243)	–	–
68	10.0221	.0469	−0.031	.488	1 (1)	1	40	.0511	−0.510	(.306)	–	1
69	10.0121	.0680	−1.664	.064	–	1	11	.0359	1.218	.126	–	1
70	10.0122	.0454	−0.627	.267	–	2	40	.0463	2.143	.019	2	3
79	10.0329	.0394	0.872	(.204)	–	–	9	.0666	0.535	.303	–	1
84	10.0175	.0516	−0.375	.358	–	1	10	.0566	0.965	.180	–	1
91	10.0395	.0566	0.743	(.235)	–	–	16	.0383	−0.729	(.239)	–	–
93	10.0045	.0440	−1.221	.116	–	2	31	.0466	2.075	.023	–	3
94	9.9841	.0467	−2.350	.013	1	3	30	.0462	1.606	.060	–	3
All	10.0190	.0503	−3.473	3×10^{-4}	10 (2)	49	1024	.0489	4.581	3×10^{-6}	13 (3)	53

* The *t*-score calculation for paired data replaces the usual *z*-score in these experiments, since an empirical baseline distribution is used as a reference in lieu of a theoretical distribution. For this calculation, an empirical standard deviation (S.D.) of the differences of the means is also required.

** The ΔPK *t*-score is calculated for the difference between the PK^+ and PK^- data, unlike the REG data where ΔPK reflects the combined effect in direction of intention.

TABLE C

Data Summary—Operator 10

Expt.	No. of Trials per Run	No. of Samples per Trial	Location	No. of Series	Intention	No. of Trials	Mean	t-Score	Prob.	No. of Series $p < .05$	No. of Series $p < .5$
REG	50	200	local	15	BL	59,100	100.011	0.388	.349	–	5
					PK^+	55,100	100.082	2.723	.003	3	12
					PK^-	55,050	99.896	−3.470	3×10^{-4}	4 (1)	13
					ΔPK	110,150		4.379	6×10^{-6}	4	14
	50	2,000	local	9	BL	20,000	999.960	−0.257	.399	–	5
					PK^+	22,350	1000.116	0.781	.217	2 (1)	6
					PK^-	22,950	999.612	−2.656	.004	3 (1)	8
					ΔPK	45,300		2.439	.007	2	8
	50	100	local	5	BL	5,250	50.028	0.406	.342	–	1
					PK^+	12,350	50.018	0.393	.347	–	3
					PK^-	13,300	49.980	−0.474	.318	–	4
					ΔPK	25,650		0.614	.270	–	4
	50	20	local	6	BL	15,000	10.010	0.460	.323	–	3
					PK^+	15,300	10.014	0.774	.219	1	3
					PK^-	14,700	10.020	1.077	(.141)	– (1)	2
					ΔPK	30,000		−0.205	(.419)	–	3
	1,000	200	local	7	BL	21,000	100.019	0.383	.351	–	3
					PK^+	21,000	100.139	2.852	.002	2	7
					PK^-	21,000	99.905	−1.958	.025	2 (1)	5
					ΔPK	42,000		3.401	3×10^{-4}	2	6
	1,000	200	remote	13	BL	39,000	100.005	0.139	.445	–	6
					PK^+	39,000	100.078	2.183	.015	3 (1)	9
					PK^-	39,000	100.018	0.500	(.309)	–	5
					ΔPK	78,000		1.190	.117		9

Data Summary—Operator 10—(Continued)

Expt.	No. of Trials per Run	No. of Samples per Trial	Location	No. of Series	Intention	No. of Trials	Mean	*t*-Score	Prob.	No. of Series $p < .05$	No. of Series $p < .5$
Pseudo	50	200	local	9	BL	22,500	99.957	−0.903	.183	–	3
					PK^+	23,000	100.059	1.275	.101	2	4
					PK^-	22,000	99.845	−3.263	6×10^{-4}	2	8
					ΔPK	45,000		3.193	7×10^{-4}	3	7
	50	2,000	local	5	BL	12,500	999.875	−0.629	.265	–	4
					PK^+	12,000	1000.060	0.290	.386	–	3
					PK^-	13,000	999.912	−0.450	.326	1	2
					ΔPK	25,000		0.525	.300	1	2
	1,000	200	local	5	BL	15,000	100.031	0.534	.297	–	1
					PK^+	15,000	100.112	1.974	.024	1	4
					PK^-	15,000	100.074	1.268	(.102)	– (2)	2
					ΔPK	30,000		0.499	.309	1	2
RMC	9,000	~ 40	local	17	BL	270 runs	10.033				
					PK^+	270 runs	10.030	−0.987	(.162)	– (2)	6
					PK^-	270 runs	10.020	−4.076	3×10^{-5}	6	14
					ΔPK	270 runs		3.235	7×10^{-4}	7 (1)	11
	9,000	~ 40	remote	5	BL	50 runs	10.013				
					PK^+	50 runs	10.019	0.754	.227	1	3
					PK^-	50 runs	10.012	−0.115	.455	–	2
					ΔPK	50 runs		0.928	.179	–	4

TABLE D

Pseudo-REG Data Summary by Operator

		PK⁺						PK⁻					
Opr.	No. of Series	No. of Trials	Mean	z-Score	Prob.	No. of Series $p < .05$	No. of Series $p < .5$	No. of Trials	Mean	z-Score	Prob.	No. of Series $p < .05$	No. of Series $p < .5$
10	9	23,000	100.059	1.269	.102	2	4	22,000	99.845	−3.249	.001	2	8
14	1	2,300	100.339	2.300	.011	1	1	2,700	100.197	1.448	(.074)	–	–
15	1	2,500	99.788	−1.499	(.067)	–	–	2,500	99.802	−1.397	.081	–	1
16	1	2,500	100.385	2.724	.003	1	1	2,500	100.122	0.860	(.195)	–	–
41	2	5,600	99.928	−0.758	(.224)	–	–	4,400	99.934	−0.616	.269	–	1
55	9	23,400	99.988	−0.251	(.401)	–	3	21,600	99.986	−0.293	.385	1	4
70	2	5,450	99.933	−0.695	(.243)	– (1)	1	4,550	99.921	−0.751	.226	1	1
80	1	2,500	100.082	0.583	.280	–	1	2,500	99.984	−0.116	.454	–	1
88	1	2,500	99.953	−0.334	(.369)	–	–	2,500	99.748	−1.782	.037	1	1
94	2	5,400	100.208	2.161	.015	1	2	4,600	99.978	−0.215	.415	–	1
All	29	75,150	100.037	1.418	.078	5 (1)	13	69,850	99.931	−2.564	.005	5	18

		Baseline						ΔPK				
Opr.	No. of Series	No. of Trials	Mean	z-Score	Prob.	No. of Series $p < .05$	No. of Series Mean > 100	No. of Trials	z-Score	Prob.	No. of Series $p < .05$	No. of Series $p < .5$
10	9	22,500	99.957	−0.904	.183	–	3	45,000	3.179	.001	3	7
14	1	2,500	100.019	0.133	.447	–	1	5,000	0.496	.310	–	1
15	1	2,500	100.183	1.293	.098	–	1	5,000	−0.072	(.471)	–	–
16	1	2,500	99.875	−0.882	.189	–	0	5,000	1.318	.094	–	1
41	2	5,000	100.138	1.384	.083	–	2	10,000	−0.158	(.437)	–	1
55	9	22,500	99.939	−1.291	.098	–	2	45,000	0.022	.491	1 (1)	5
70	2	5,000	100.139	1.392	.082	–	2	10,000	−0.007	(.497)	1 (1)	1
80	1	2,500	100.089	0.631	.264	–	1	5,000	0.494	.311	–	1
88	1	2,500	99.886	−0.809	.209	–	0	5,000	1.024	.153	–	1
94	2	5,000	99.717	−2.834	.002	1	0	10,000	1.734	.041	1	2
All	29	72,500	99.969	−1.170	.121	1	12	145,000	2.801	.003	6 (2)	20

NOTES

Preamble

1. Niels Bohr, *Atomic Theory and the Description of Nature* (Cambridge: The University Press, 1961), p. 119.

SECTION I

1. Mircea Eliade, *Myths, Dreams, and Mysteries*, trans. P. Mairet (New York: Harper & Row, 1960), p. 13.

Chapter 3

2. Hermes Trismegistus, *Emerald Tablet*, in J. Ruska, *Tabula Smaragdina* (Heidelberg: Carl Winter's Universitätsbuchhandlung, 1926), p. 2.
3. Roger Bacon, in D. E. Smith, "The Place of Roger Bacon in the History of Mathematics," in A. G. Little, ed., *Roger Bacon: Essays* (Oxford: Clarendon Press, 1914), p. 178. (Quoted in Will Durant, *The Age of Faith*. New York: Simon and Schuster, 1950, p. 1010.)
4. Nicholas Copernicus. (Quoted in J. Bronowski, *The Ascent of Man* [Boston and Toronto: Little, Brown and Co., 1973], pp. 196–97.)
5. William Gilbert, *De Magnete*, Book V, chapter 12, trans. P. Fleury (Mottelay, London, 1893). (Quoted in H. Kearney, *Science and Change 1500–1700* [New York and Toronto: McGraw-Hill, World University Library, 1971], pp. 110–12.)
6. John M. Keynes, "Newton, the Man," in G. Keynes, ed., *Essays in Biography* (New York: W. W. Norton & Co., 1963), p. 311.

7. D. Kubrin, "Newton's Inside Out! Magic, Class Struggle, and the Rise of Mechanism in the West," in H. Woolf, ed., *The Analytic Spirit: Essays in the History of Science* (Ithaca and London: Cornell University Press, 1981), p. 113.
8. Francis Bacon, *Novum Organum,* Part I. (Quoted in F. S. Taylor, *Science Past and Present* [London and Toronto: William Heinemann, 1945], p. 86.)
9. Robert Hooke. (Quoted in Henry Lyons, *The Royal Society 1660–1940* [Cambridge: The University Press, 1944], p. 41.)
10. Francis Bacon. (In D. P. Walker, "Francis Bacon and *Spiritus,*" A. G. Debus, ed., *Science, Medicine and Society in the Renaissance* [New York: Neale Watson Academic Publications, Inc. (Science History Publications), 1972], p. 127.)
11. Robert Hooke, in R. Waller, ed., *The posthumous works of Robert Hooke, M.D., S.R.S., . . . containing his Cutlerian lectures and other discourses, read at the meetings of the illustrious Royal Society* (London: Smith and Walford [Printers to the Royal Society], 1705), p. 147. (Quoted in B. R. Singer, "Robert Hooke on Memory, Association, and Time Perception," in R. V. Jones and W. D. M. Paton, eds., *Notes and Records of the Royal Society of London* 31, no. 1 [July 1976], pp. 123–24.)
12. Robert Boyle, *Works,* vol. 1, p. cxxx. (Quoted in L. T. More, *Isaac Newton: A Biography* [New York: Dover Publications, 1962], p. 163.)
13. F. A. Yates, *The Rosicrucian Enlightenment* (Boulder, Colo.: Shambhala, 1978).
14. Henry More, *An Antidote Against Atheism.* (Quoted in Editor's Introduction, C. O. Parsons, ed., J. Glanvill, *Saducismus Triumphatus* [Gainesville, Fla.: Scholars' Facsimiles and Reprints, 1966], p. viii.)
15. Immanuel Kant, *A Critique of Pure Reason,* trans. N. K. Smith (London: Macmillan and Co., 1956), p. 125. (Quoted in W. and A. Durant, *Rousseau and Revolution* [New York: Simon and Schuster, 1967], p. 537.)
16. Henri Bergson, *Creative Evolution,* trans. A. Mitchell (New York: Henry Holt and Co., 1923), p. 179.
17. *Ibid.*, p. 263.
18. William James, *Some Problems of Philosophy: A Beginning of an Introduction to Philosophy* (New York: Longmans, Green, & Co., 1911), p. 50.

19. R. Haynes, *Philosopher King: The Humanist Pope Benedict XIV* (London: Weidenfeld and Nicholson, 1970), pp. 100–103.
20. Emanuel Swedenborg, in G. Trobridge, *Swedenborg: Life and Teaching* (New York: Swedenborg Foundation, Inc., 1976), p. 134.
21. Israel Regardie, *The Philosopher's Stone* (St. Paul, Minn.: Llewellyn Publications, 1978), p. 92–93.
22. Carl G. Jung, "Synchronicity: An Acausal Connecting Principle," in C. G. Jung and W. Pauli, *The Interpretation of Nature and the Psyche*, trans. R. F. C. Hull (New York: Pantheon Books, Bollingen Series LI, 1955), (Copyright Princeton University Press), pp.5–146.

Chapter 4

23. William James, "Review of Human Personality and Its Survival of Bodily Death by F. W. H. Myers." *Proceedings of the Society for Psychical Research* 18 (1903), p. 23.
24. F. W. H. Myers, *Human Personality and Its Survival of Bodily Death*, L. H. Myers, ed. (London and Bombay: Longmans, Green, & Co., 1907), pp. xi, 9.
25. *Ibid.*, p. 73.
26. Charles Richet, "La suggestion mentale et la calcul des probabilités." *Revue Philosophique* 18 (1884), pp. 608–74.
27. Sigmund Freud, "Dreams and Telepathy," in *Studies in Parapsychology*, P. Rieff, ed. (New York: Collier Books, 1963), p. 88.
28. Ernest Jones, *The Life and Work of Sigmund Freud*, vol. 3 (New York: Basic Books, 1957), p. 392.
29. J. B. Rhine, "Psi and psychology: Conflict and solution," *Journal of Parapsychology* 32 (1968), pp. 101–28.
30. Wolfgang Pauli, *Aufsätze und Vorträge über Physik und Erkenntnistheorie* (Essays and Lectures on Physics and Epistemology) (Braunschweig: Friedr. Vieweg und Sohn, 1961), p. 124.

Chapter 6

31. Niels Bohr, *Atomic Theory and the Description of Nature* (Cambridge: The University Press, 1961).
32. Wolfgang Pauli, "The Influence of Archetypal Ideas on the Scientific Theories of Kepler," in C. G. Jung and W. Pauli, *The Interpretation of Nature and the Psyche*, trans. P. Silz (New York:

Pantheon Books, Bollingen Series LI, 1955), pp. 149–240. (See also W. Pauli, *Aufsätze und Vorträge über Physik und Erkenntnistheorie* [Braunschweig: Friedr. Vieweg und Sohn], 1961.)

33. Erwin Schrödinger, *My View of the World*, trans. C. Hastings (Cambridge: The University Press, 1964). (See also E. Schrödinger, *Nature and the Greeks* [Cambridge: The University Press, 1954].)
34. Werner Heisenberg, *Physics and Philosophy* (New York: Harper & Row, Harper Torchbooks, 1958). (See also W. Heisenberg, *Physics and Beyond*, trans. A. J. Pomerans [New York: Harper & Row, 1972].)
35. Pascual Jordan, "Parapsychological implications of research in atomic physics," *International Journal of Parapsychology* 2, no. 4 (1960), pp. 5–16.
36. Eugene Wigner, *Symmetries and Reflections* (Bloomington and London: Indiana University Press, 1967).
37. Carl Friedrich von Weizsäcker, *The Unity of Nature*, trans. F. J. Zucker (New York: Farrar, Straus, Giroux, 1980; Munich: Carl Honser Verlag, 1971).
38. Albert Einstein, *Out of My Later Years* rev. reprint ed. (Secaucus, N.J.: The Citadel Press, 1956). (See also P. A. Schilpp, ed., *Albert Einstein: Philosopher-Scientist* [Evanston, Ill.: The Library of Living Philosophers], 1949.)
39. Max Planck, *A Survey of Physical Theory*, trans. R. Jones and D. H. Williams (New York: Dover Publications, 1960), p. 53.
40. Max Planck, *The Universe in the Light of Modern Physics*, trans. W. H. Johnston (New York: W. W. Norton & Co., 1931), pp. 113–14.
41. Max Planck, *Where Is Science Going?*, trans. J. Murphy (New York: W. W. Norton & Co., 1932), pp. 159–60.
42. Erwin Schrödinger, "Mind and Matter," in *What Is Life? and Mind and Matter* (Cambridge: The University Press, 1967), p. 137.
43. Erwin Schrödinger, *My View of the World* (Cambridge: The University Press, 1964), p. 40.
44. Erwin Schrödinger, "What Is Real?" in *My View of the World* (Cambridge: The University Press, 1964), p. 94.
45. Erwin Schrödinger, "Mind and Matter," in *What Is Life? and Mind and Matter* (Cambridge: The University Press, 1967), p. 131.
46. Louis de Broglie, *New Perspectives in Physics*, trans. A. J. Pomerans (New York: Basic Books, 1962), p. 221.

47. Louis de Broglie, *Matter and Light: The New Physics*, trans. W. H. Johnston (New York: W. W. Norton & Co., 1939), pp. 281–92.
48. R. G. Jahn and B. J. Dunne, "On the Quantum Mechanics of Consciousness with Application to Anomalous Phenomena. Appendix B: Collected Thoughts on the Role of Consciousness in the Physical Representation of Reality." Technical Note PEAR 83005.1B, Princeton Engineering Anomalies Research, Princeton University, School of Engineering/Applied Science, 1984.

Chapter 9

49. Albert Einstein. (In L. Barnett, *The Universe and Dr. Einstein*, rev. ed. [New York: William Morrow Bantam Books, 1979], p. 108.)

SECTION II

1. Friedrich von Schiller, *Wallensteins Tod* (The Death of Wallenstein), Act II, Scene 3. W. Witte, ed. (Oxford: Basil Blackwell, 1952), p. 186.

Chapter 1

2. Helmut Schmidt, "Anomalous Prediction of Quantum Processes by Some Human Subjects," Boeing Scientific Research Laboratories Document #D1-82-0821, February 1969. (See also H. Schmidt, "A PK test with electronic equipment," *Journal of Parapsychology* 34, no. 3 [1970], pp. 175–81.)

Chapter 3

3. R. D. Nelson, B. J. Dunne, R. G. Jahn, "An REG Experiment with Large Data Base Capability, III: Operator Related Anomalies." Technical Note PEAR 84003, Princeton Engineering Anomalies Research, Princeton University, School of Engineering/Applied Science, 1984.

Chapter 4

4. G. Spencer Brown, "Statistical significance in psychical research," *Nature* 172 (July 25, 1953), pp. 154–56.

5. Abner Shimony, "Metaphysical problems in the foundations of quantum mechanics," *International Philosophical Quarterly* 18 (1978), pp. 3–17.

Chapter 6

6. Francis Galton, *Natural Inheritance* (New York and London: Macmillan and Co., 1894), pp. 63–65.
7. R. D. Nelson, B. J. Dunne, R. G. Jahn, "Operator Related Anomalies in a Random Mechanical Cascade Experiment." Technical Note PEAR 88001, Princeton Engineering Anomalies Research, Princeton University, School of Engineering/Applied Science, 1988.

SECTION III

1. Franz Hartmann, *Paracelsus: Life and Prophecies* (Blauvelt, N.Y.: Rudolf Steiner Publications, 1973), p. 105.

Chapter 1

2. H. E. Puthoff and R. Targ, "A perceptual channel for information transfer over kilometer distances: Historical perspective and recent research," *Proceedings IEEE* 64 (1976), pp. 329–54.

Chapter 2

3. B. J. Dunne, R. G. Jahn, R. D. Nelson, "Precognitive Remote Perception." Technical Notes PEAR 83003 and 89001, Princeton Engineering Anomalies Research, Princeton University, School of Engineering/Applied Science, 1983 and 1989.

SECTION IV

1. Herman Melville, *Moby-Dick: An Authoritative Text, Reviews and Letters by Melville, Analogues and Sources, Criticism.* H. Hayford and H. Parker, eds. (New York and London: W. W. Norton & Co., 1967), p. 264.

Chapter 2

2. Francis Bacon, "Idols of perception," in *Novum Organum*. (Quoted in G. B. Levitas, ed., *The World of Psychology*, vol. I [New York: George Braziller, 1963], p. 161.)
3. James Jeans, *The Mysterious Universe* (New York: Macmillan Co.; Cambridge: The University Press, 1948), pp. 166, 186.
4. Sigmund Freud, *The Future of an Illusion* (London, 1943), p. 97. (Quoted in Editor's Introduction, S. Freud, *General Psychological Theory* [New York: Collier Books, 1963] p. 9.)
5. Arthur Schopenhauer, *The World as Will and Representation*, vol. II, trans. E. F. J. Payne (New York: Dover Publications, 1966), pp. 15–16.
6. Albert Einstein, in P. A. Schilpp, ed., *Albert Einstein: Philosopher-Scientist* (Evanston, Ill.: The Library of Living Philosophers, 1949), pp. 175–76.
7. *Ibid.*, p. 13.

Chapter 3

8. Werner Heisenberg, *Physics and Philosophy* (New York: Harper & Row, Harper Torchbooks, 1958), p. 81.

Chapter 5

9. William James, *Some Problems of Philosophy: A Beginning of an Introduction to Philosophy* (New York: Longmans, Green, & Co., 1911), p. 185.
10. William James, *Psychology: Briefer Course* (New York: Collier Books, 1962), p. 211.
11. E. Pöppel, "Oscillations as possible basis for time perception," in J. T. Fraser, F. C. Maker, G. H. Müller, eds., *The Study of Time: Proceedings of the First Conference of the International Society for the Study of Time, Oberwolfach (Black Forest, West Germany, 1969)* (New York, Heidelberg, Berlin: Springer-Verlag, 1972), p. 235.
12. Sigmund Freud, *The Neuro-Psychoses of Defense* (London: Collected Papers, vol. 1, 1984). (Quoted in Editor's Introduction, J. Strachey, ed. and trans., *S. Freud: The Interpretation of Dreams* [New York: Avon Books, 1965], p. xvi.)
13. Hermann Weyl, *Mind and Nature* (Philadelphia: University of Pennsylvania Press, 1934), pp. 19–20.

14. Carl G. Jung, "On the Nature of the Psyche," *Collected Works*, vol. 8 (New York: Pantheon Books, 1960), pp. 233–34.

Chapter 6

15. Erwin Schrödinger, "Mind and Matter" in *What Is Life? and Mind and Matter* (Cambridge: The University Press, 1967), pp. 127–28.

Chapter 8

16. J. W. Goethe, in F. Ungar, ed., *Practical Wisdom: A Treasury of Aphorisms and Reflections from the German* (New York: Frederick Ungar Publishing Co., 1977), p. 131.
17. Niels Bohr, *Atomic Theory and the Description of Nature* (Cambridge: The University Press, 1961), pp. 20, 24.
18. Werner Heisenberg, *Physics and Beyond* (New York: Harper & Row, Harper Torchbooks, 1972), p. 114.
19. *Ibid.*, p. 115
20. Werner Heisenberg, *Physics and Philosophy: The Revolution in Modern Physics* (New York: Harper & Row, Harper Torchbooks, 1958), p. 179.
21. Wolfgang Pauli, "The Influence of Archetypal Ideas on the Scientific Theories of Kepler," in C. G. Jung and W. Pauli, *The Interpretation of Nature and the Psyche* (New York: Pantheon Books, Bollingen Series LI, 1955), pp. 208, 210.
22. Benjamin Franklin, *Poor Richard's Almanack* (Mt. Vernon, Peter Pauper Press, 1939).
23. J. W. Goethe in F. Ungar, ed., *Goethe's World View, Presented in His Reflections and Maxims* (New York: F. Ungar Publishing Co., 1963), p. 139.

Chapter 11

24. William James, *Some Problems of Philosophy: A Beginning of an Introduction to Philosophy* (New York: Longmans, Green, & Co., 1911), pp. 50–51.
25. Albert Einstein, in *Albert Einstein, the Human Side: New Glimpses from His Archives*, H. Dukas and B. Hoffmann, eds. (Princeton, N.J.: Princeton University Press, 1981), p. 38.

SECTION V

1. Louis de Broglie, "The Role of the Engineer in the Age of Science," in *New Perspectives in Physics*, trans. A. J. Pomerans (New York: Basic Books, 1962), p. 213.

Chapter 5

2. Arthur Eddington, *The Nature of the Physical World* (Ann Arbor: The University of Michigan Press, 1978), pp. 313, 314–15.

Chapter 6

3. A. Einstein, B. Podolsky, and N. Rosen, "Can quantum mechanical description of reality be considered complete?" *Physical Review* 47, no. 777 (1935). (See also J. S. Bell, "Quantum interconnectedness," *Physics* 1, no. 195 [1965].)
4. A. Aspect, J. Dalibard, and G. Roger, "Experimental test of Bell's Inequalities using time-varying analyzers," *Physical Review Letters* 49 (1982), p. 1804.
5. Erwin Schrödinger, *Nature and the Greeks* (Cambridge: The University Press, 1945), p. 91.
6. Erwin Schrödinger, *What Is Life?* (Cambridge: The University Press; New York: Macmillan Co., 1945), p. 88.
7. Arthur Eddington, *Relativity Theory of Protons and Electrons* (New York: Macmillan Co., and Cambridge: The University Press, 1936), p. 329.
8. Werner Heisenberg, "The nature of elementary particles," *Physics Today* 29 (1976), pp. 32–39.
9. Sidney D. Drell, "When is a particle?" *Physics Today* 31 (1978), pp. 23–32.
10. Fritjof Capra, Paper LBL-796, Lawrence Berkeley Laboratories, University of California, Berkeley, 1978.

Chapter 7

11. William James, "Psychical Research," in *The Will to Believe and Other Essays in Popular Philosophy and Human Immortality* (New York: Dover Publications, Inc., 1956), p. 327.
12. Henri Bergson, *An Introduction to Metaphysics*, second (rev.) ed.,

trans. T. E. Hulme (Indianapolis and New York: The Bobbs-Merrill Co., The Liberal Arts Press, 1955), pp. 53–54.
13. James Jeans, *Physics and Philosophy* (Cambridge: The University Press, 1943), p. 200.
14. Pierre Teilhard de Chardin, *The Phenomenon of Man,* trans. B. Wall (New York: Harper & Row, 1965), p. 36.
15. C. M. Pleass and N. D. Dey, "The use of cultures of marine microalgae as biotransducers in the scientific study of the energy fields of living organisms," College of Marine Studies, University of Delaware (personal communication).
16. I. Prigogine and I. Stengers, *Order Out of Chaos: Man's New Dialogue with Nature* (New York: Bantam Books, 1984).
17. Rupert Sheldrake, *A New Science of Life: The Hypothesis of Formative Causation* (Los Angeles: J. P. Tarcher, 1981).

Chapter 8

18. Albert Einstein, *Out of My Later Years,* rev. reprint ed. (Secaucus, N.J.: The Citadel Press, 1956), p. 26.
19. Max Planck, in F. Ungar, ed., *Practical Wisdom: A Treasury of Aphorisms and Reflections from the German* (New York: Frederick Ungar Publishing Co., 1977), p. 19.

Chapter 9

20. Pierre Teilhard de Chardin, *On Love* (New York: Harper & Row, 1967), pp. 33–34. (From P. Teilhard de Chardin, "The Evolution of Chastity.")

Chapter 10

21. Arthur Eddington, *The Nature of the Physical World* (Ann Arbor: The University of Michigan Press, 1978), p. 241.
22. Arthur Eddington, *Space, Time and Gravitation* (Cambridge: The University Press, 1978), pp. 200–201.

SOURCES FOR SIDEBAR QUOTATIONS

page number

15 Israel Regardie, *The Middle Pillar*. St. Paul, Minn.: Llewellyn Publications, 1978, p. 19.

20 James Jeans, *Physics and Philosophy*. Cambridge: The University Press, 1943, p. 204.

24 Johannes Kepler, *Harmonices mundi*, Book V (Frisch, vol. V, p. 223) in W. Pauli, "The Influence of Archetypal Ideas on Kepler's Theories." In C. G. Jung and W. Pauli, *The Interpretation of Nature and the Psyche*. (Tr. R. F. C. Hull.) New York: Pantheon Books (Bollingen Series LI), 1955, p. 161.

48 J. B. Rhine, *The Reach of the Mind*. New York: William Sloane Associates, Inc., 1962, p. 153.

51 Stephen Jay Gould, "Bligh's Bounty," *Natural History*, vol. 94, no. 9, September 1985, pp. 4, 6–7.

57 Max Planck, *Where Is Science Going?* (Tr. J. Murphy.) New York: W. W. Norton & Co., 1932, p. 117.

58 Niels Bohr, *Atomic Theory and the Description of Nature*. Cambridge: The University Press, 1961, p. 116.

195 Hazrat Inayat Khan, *The Complete Sayings of Hazrat Inayat Khan*. New Lebanon, N.Y.: Sufi Order Publications, 1978, p. 42.

205 Jonas Salk, *The Survival of the Wisest.* New York: Harper & Row, 1973, p. 45.

207 Werner Heisenberg, *Physics and Beyond.* New York: Harper & Row, 1971, p. 101.

211 James Jeans, *Physics and Philosophy.* Cambridge: The University Press, 1943, p. 204.

218 Immanuel Kant, *Critique of Pure Reason.* (Tr. J. M. D. Meiklejohn.) London: Henry G. Bohn, 1855, p. 118.

222 Arthur Eddington, *The Philosophy of Physical Science.* New York: The Macmillan Co., 1939, p. 129.

224 Louis de Broglie, *Physics and Microphysics.* (Tr. M. Davidson.) New York: Pantheon Books, 1955, p. 131.

227 Arthur Eddington, *Space, Time, and Gravitation.* Cambridge: The University Press, 1978, p. 150.

228 Niels Bohr, *Atomic Theory and the Description of Nature.* Cambridge: The University Press, 1961, p. 19.

239 Werner Heisenberg, *Physics and Philosophy.* New York: Harper & Row (Harper Torchbooks), 1958, p. 106.

240 P. D. Ouspensky, *Tertium Organum.* New York: Random House (Vintage Books), 1981, p. 290.

247 Alfred North Whitehead, *Nature and Life.* Chicago: University of Chicago Press, 1934, p. 46.

255 Luther Burbank, in W. S. Harwood, "A wonder-worker of science: An authoritative account of Luther Burbank's unique work in creating new forms of plant life." *The Century Magazine,* no. 69, 1904–1905, p. 838.

258 Erwin Schrödinger, *What Is Life? and Mind and Matter.* Cambridge: The University Press, 1960, p. 176.

260 Arthur Eddington, *The Philosophy of Physical Science.* New York: Macmillan, 1939, p. 128.

264 Louis de Broglie, *Matter and Light: The New Physics.* (Tr. W. H. Johnston.) New York: W. W. Norton, 1939, p. 280.

284 Albert Einstein, in W. Heisenberg, *Physics and Beyond.* New York: Harper & Row, 1971, p. 63.

284 Charlie Chan, "Charlie Chan in Egypt," 1935. (Quoted in S. J. Gould, "Bligh's Bounty," *Natural History,* vol. 94, no. 9, September 1985, p. 10.)

285 Carl Jung, "Analytical Psychology and Education." *Collected Works,* vol. 17. New York: Pantheon Books, 1954, p. 89.

292 Goethe, in F. Ungar, *Practical Wisdom: A Treasury of Aphorisms and Reflections from the German.* New York: Frederick Ungar Publishing Co., 1977, p. 25.

293 Rollo May, *Love and Will.* New York: Dell Publishing Co. (A Delta Book), 1969, p. 325.

298 Carl Friedrich von Weizsäcker, *The Unity of Nature.* New York: Farrar, Straus, Giroux, Inc., 1980, pp. 278–79.

301 Edgar D. Mitchell, "From Outer Space to Inner Space . . ." in E. D. Mitchell, *Psychic Exploration: A Challenge for Science,* J. White, ed. New York: G. P. Putnam's Sons, 1974, pp. 25, 43–44.

306 Albert Einstein, in P. A. Schilpp, ed., *Albert Einstein: Philosopher-Scientist.* Evanston, Ill.: The Library of Living Philosophers (George Banta Publishing Co., Menasha, Wisc.), 1949, p. 49.

308 Arthur Eddington, *The Nature of the Physical World.* Ann Arbor, Mich.: The University of Michigan Press, 1978, pp. 344–45.

310 Lao Tsu, *Tao Te Ching.* (Tr. G-F. Feng and J. English.) New York: Random House (Vintage Books), 1972, Chapter 63.

312 Pierre Teilhard de Chardin, *L'Apparition de l'homme,* pp. 179–82. (Quoted in C. E. Raven, *Teilhard de Chardin: Scientist and Seer.* New York and Evanston: Harper & Row, 1962, p. 40.)

324 Alfred Russel Wallace, *Man's Place in the Universe: A Study of the Results of Scientific Research in Relation to the Unity or Plurality of Worlds.* New York: McClure Phillips and

Company, 1903. (Quoted in S. J. Gould, *The Flamingo's Smile: Reflections in Natural History*. New York and London: W. W. Norton & Co., 1985, p. 397.)

326 Carl Jung, *Memories, Dreams, Reflections*. New York: Vintage Books, 1965, p. 137.

333 P. D. Ouspensky, *Tertium Organum*. New York: Random House (Vintage Books), 1981, p. 194.

335 Alan Watts, *Uncarved Block, Unbleached Silk: The Mystery of Life*. New York: A & W Visual Library, 1978.

337 Arthur Eddington, *Space, Time and Gravitation*. Cambridge: The University Press, 1978, p. 198.

338 Lincoln Barnett, *The Universe and Dr. Einstein*. New York: William Morrow (Bantam Books), 1979, p. 35.

343 F. W. H. Myers, *Human Personality and Its Survival of Bodily Death*. London and Bombay: Longmans, Green & Co., 1907, p. 344.

ILLUSTRATION CREDITS

SECTION ONE

The Vectors

P. 1, Culver Pictures; *p. 4,* Courtesy of The New York Public Library (from Michael Majer, *Atalanta Fugiens,* 1687); *p. 6,* Courtesy of the New York Public Library (from Mainz Breydenback, *Travels in the Holy Land,* 1486); *p. 7,* Culver Pictures; *p. 11,* Dr. Fabrizio Mori, Nannini, Rome; *p. 12,* The Bettmann Archive; *p. 14,* Culver Pictures; *p. 16,* Samuel Weiser, Inc./U.S. Games Systems, Inc.; *p. 18,* Culver Pictures; *p. 21,* Picture Collection, The New York Public Library; *p. 23,* Österreichische Nationalbibliothek, Vienna; *p. 26,* Vadiana Library, St. Gall; *p. 29,* Culver Pictures; *p. 33,* Culver Pictures; *p. 36,* Bridgeman/Art Resource/British Museum; *p. 39,* Courtesy of Charles Scribner's Sons; *p. 40* top, Culver Pictures; *p. 40* bottom, Courtesy of The New York Public Library (from R. G. Medhurst, *Crookes and the Spirit World,* 1972); *p. 43* both Courtesy of Robert G. Jahn and Brenda J. Dunne; *p. 45,* Courtesy of The New York Public Library (from Joseph Glanvil, *Sadducismus Triumphatus,* 1681): *p. 50, Harper's Magazine,* December 1876; *p. 53,* Culver Pictures; *p. 55,* The Bettmann Archive; *p. 56,* Courtesy of Woldemar von Jaskowsky; *p. 59* top, Courtesy of The New York Public Library (From Max Börn, *Atomic Physics*); *p. 59* bottom, Courtesy of The New York Public Library (from Lincoln Barnett, The Universe and Dr. Einstein); *p. 65,* Ken Love/Black Star; *p. 68,* Culver Pictures (Sir Isaac Newton); *p. 78,* Four By Five; *p. 79,* Michael L. Freeman with Robert G. Jahn and Brenda J. Dunne; *p. 81,* Michael L. Freeman with Robert G. Jahn and Brenda J. Dunne; *p. 82,* Courtesy of The New York Public Library (from Heinrich Khunrath, *Amphitheatrum Aeternae Sapientiae,* 1609)

SECTION TWO

Man/Machine Margins

P. 85, Culver Pictures; *p. 88,* Culver Pictures; *p. 94* top, Donald Neiler; *p. 125,* Courtesy of The New York Public Library (from Francis Galton, *Natural Inheritance,* 1884); *p. 126,* Karen Halverson

SECTION FOUR

The Waves of Consciousness

P. 193, Culver Pictures/Courtesy of the Rockwell Kent Legacies; *p. 197* top, Oswald White Bear Fredericks, Courtesy of Viking Press (Hopi labyrinth); *p. 197* center, from Jill Purce, *The Mystic Spiral,* 1980 (labyrinth on island of Weir, Finland); *p. 198,* Courtesy of *Parabola Magazine,* 1979 (Cretan labyrinth); *p. 199,* Courtesy of *Parabola Magazine,* 1979 (labyrinth in Chartres Cathedral); *p. 201,* from Jill Purce, *The Mystic Spiral,* 1980 (labyrinth in Toussaints Abbey); *p. 204,* Michael L. Freeman with Robert C. Jahn and Brenda J. Dunne; *p. 206,* Culver Pictures; *p. 209,* Robert G. Jahn; *p. 210* top, Donald Neiler; *p. 210* bottom, Courtesy of Dr. G. L. Germer; *p. 213* top, Culver Pictures; *p. 213* center, Culver Pictures; *p. 214,* Courtesy of The New York Public Library (from Henry DeWolf Smyth and Charles Wilbur Ufford, *Matter, Motion and Electricity,* 1939); *p. 215,* Culver Pictures; *p. 216,* Eugene Anthony/Black Star (open water waves); *p. 219,* Culver Pictures; *p. 226,* Science Source/Photo Researchers, Inc.; *p. 229,* Österreichische Nationalbibliothek, Vienna; *p. 231,* Mary Evans Picture Library/Photo Researchers, Inc.; *p. 234,* from Larousse, *World Mythology*; *p. 236,* Smithsonian Institution; *p. 243,* Michael L. Freeman with Robert G. Jahn and Brenda J. Dunne; *p. 245,* Michael L. Freeman with Robert G. Jahn and Brenda J. Dunne; *p. 246,* Michael L. Freeman with Robert G. Jahn and Brenda J. Dunne; *p. 248,* Marc Hasselnis; *p. 250,* Courtesy of The New York Public Library (from Max Börn, *Atomic Physics*); *p. 252,* George Cruikshank/Culver Pictures; *p. 254,* Michael L. Freeman with Robert G. Jahn and Brenda J. Dunne; *p. 256,* Culver Pictures; *p. 259,* Reece Winston; *p. 261,* Culver Pictures; *p. 262,* Culver Pictures; *p. 265,* A. I. P. Niels Bohr Library/Margrethe Bohr Collection; *p. 269,* © 1987 by Sidney Harris/*American Scientist Magazine*; *p. 272,* Robert P. Matthews; *p. 274,* © 20th Century Fox Film Corporation (Artoo & Deetoo with See Threepio, robots from *Star Wars*); *p. 277,* from G. W. King, *The Gnostics & Their Remains,* 1887; *p. 281,* Courtesy of The New York Public Library; *p. 284,* Giraudon/Art Resource/Bibliothèque Nationale, Paris; *p. 287,* Culver Pictures (Adam naming the animals)

SECTION FIVE

The Vectors Deflected

P. 289, Mount Wilson & Palomar Observatories, Courtesy of the California Institute of Technology; *p. 294,* Culver Pictures (web); *p. 296,* James Edward Vaughan/Black Star; *p. 302,* Courtesy of Harper & Row Publishers, Inc. (from J.A. West, *Serpent in the Sky,* 1979); *p. 305,* William Hogarth/Courtesy of the Trustees of the British Museum; *p. 307,* American Museum of Natural History; *p. 310,* Courtesy of American Heritage; *p. 316,* Martin Saint Martin/*Physics Today,* April 1985; *p. 318* left, Mount Wilson & Palomar Observatories, Courtesy of the California Institute of Technology; *p. 318* right, Lawrence Radiation Laboratory, Berkeley, California/Science Source/Photo Researchers, Inc.; *p. 320* top, Michael L. Freeman with Robert G. Jahn and Brenda J. Dunne; *p. 320* bottom, Culver Pictures; *p. 325,* Courtesy of Dr. Ronald H. Cohn, The Gorilla Foundation; *p. 327,* Culver Pictures; *p. 329,* René Magritte, Giraudon/Art Resource/Musée Nationale d'Art Moderne, Paris; *p. 330,* Michael L. Freeman with Robert G. Jahn and Brenda J. Dunne; *p. 334,* Hans Holbein, *Das Buchdruckereichen des Johannes Foben*/Öffentliche Kunstsammlung Basel, Kuntsmuseum; *p. 336,* Ellen Mogensen; *p. 339,* painting by Eitaku Kobayashi, William Sturgis Bigelow Collection, Courtesy, Museum of Fine Arts, Boston; *p. 342,* Courtesy of Dell Publishing Co.; *p. 344,* Courtesy of The New York Public Library

P. 349, Marc Hasselnis, after Mallowen

SUGGESTIONS FOR FURTHER READING

Many of the particular references designated by the numbered notes throughout the text contain additional material that may interest the general reader, and collectively these constitute a good initial reading list. Beyond these, we suggest the following sources that have proven valuable in the preparation of this book and can provide starting points for more extensive study of many of the topics we have touched on only briefly. Although these are somewhat arbitrarily assigned to the individual sections and chapters, many of them, and their bibliographies, are pertinent to several of the areas addressed.

SECTION ONE

The Vectors

1. SCIENTIFIC TWO-STEP

B. Barber, "Resistance by scientists to scientific discovery," *Science* 134 (1961): 596–602.

I. B. Cohen, *Revolution in Science.* Cambridge, Mass.: Harvard University Press (Belknap Press), 1985.

P. Feyerabend, *Against Method: Outline of an Anarchistic Theory of Knowledge.* London: Humanities Press, 1975.

R. Harré, ed., *Problems of Scientific Revolution: Progress and Obstacles to Progress in the Sciences.* Oxford: Clarendon Press, 1975.

M. B. Hesse, *The Structure of Scientific Inference.* London: Macmillan, 1974.

T. S. Kuhn, *The Structure of Scientific Revolutions.* Chicago: The University of Chicago Press, 1970.

K. Popper, *The Logic of Scientific Discovery.* London: Hutchinson, 1972.

R. Wallis, ed., *On the Margins of Science: The Social Construction of Rejected Knowledge.* Hanley, Stoke-on-Trent, England: J. H. Brookes (Printers), 1956.

2. MAN AND THE MYSTICAL

J. Campbell, *The Mythic Image.* Princeton, N.J.: Princeton University Press, 1974.

J. Campbell, ed., *The Mysteries: Papers from the Eranos Yearbooks.* New York: Pantheon Books (Bollingen Series XXX.2), 1955.

———, *Spiritual Disciplines: Papers from the Eranos Yearbooks.* New York: Pantheon Books (Bollingen Series XXX.4), 1960.

———, *The Mystic Vision: Papers from the Eranos Yearbooks.* New York: Pantheon Books (Bollingen Series XXX.6), 1969.

É. Durkheim, *The Elementary Forms of the Religious Life*, trans. J. Swain. New York: Macmillan (The Free Press), 1951.

M. Eliade, *Patterns in Comparative Religion.* New York: New American Library (Meridian), 1958.

———, *The Sacred and the Profane: The Nature of Religion*, trans. W. R. Trask. New York: Harcourt, Brace, & World, Inc. (Harvest), 1959.

M. Ferguson, *The Aquarian Conspiracy.* Los Angeles: Tarcher, 1979.

J. G. Frazer, *The New Golden Bough: A New Abridgment of the Classic Work*, T. H. Gaster, ed. New York: New American Library (Mentor Books), 1959.

P. Grimal, *Larousse World Mythology.* New York: Prometheus Press, 1965.

E. Hadingham, *Early Man and the Cosmos.* Norman, Okla.: University of Oklahoma Press, 1984.

W. W. Harman, *An Incomplete Guide to the Future.* New York: Norton, 1976.

J. Hawkes and L. Woolley, *History of Mankind:* Vol. 1, *Prehistory and the Beginnings of Civilization.* New York: Harper & Row, 1963.

W. James, *The Varieties of Religious Experience.* New York: New American Library (Mentor Books), 1958.

C. G. Jung and C. Kerényi, *Essays on a Science of Mythology: The Myth of the Divine Child and the Mysteries of Eleusis*, trans. R. F. C. Hull.

Princeton, N.J.: Princeton University Press (Bollingen Series XXII), 1973.

E. Neumann, *The Origins and History of Consciousness*, trans. R. F. C. Hull. Princeton, N.J.: Princeton University Press (Bollingen Series XLII), 1973.

P. D. Ouspensky, *Tertium Organum: A Key to the Enigmas of the World*, trans. E. Kadloubovsky and P. D. Ouspensky. New York: Random House (Vintage Books), 1981.

E. H. Pagels, *The Gnostic Gospels*. New York: Vintage Books, 1981.

R. A. Schwaller de Lubicz, *Sacred Science: The King of Pharaonic Theocracy*, trans. A. and G. Vandenbroeck. New York: Inner Traditions International, 1982.

R. F. Spencer and J. D. Jennings, *The Native Americans*. New York: Harper & Row, 1977.

W. T. Stace, ed., *The Teachings of the Mystics: Being Selections from the Great Mystics and Mystical Writings of the World*. New York: New American Library (Mentor Books), 1980.

C. T. Tart, ed., *Transpersonal Psychologies*. New York: Harper & Row, 1975.

A. W. Watts, *The Two Hands of God: The Myths of Polarity*. New York: Macmillan (Collier Books), 1963.

H. Zimmer, *Myths and Symbols in Indian Art and Civilization*, J. Campbell, ed. Princeton, N.J.: Princeton University Press (Bollingen Series VI), 1974.

3. SCHOLARLY STREAM

H. Bergson, *Time and Free Will: An Essay on the Immediate Data of Consciousness*. London: Allen and Unwin, 1921.

———, *An Introduction to Metaphysics*, trans. T. E. Hulme. Indianapolis: Bobbs-Merrill (Liberal Arts Press), 1955.

M. Boas, *The Scientific Renaissance 1450–1630*. New York: Harper & Row (Harper Torchbooks: Science Library), 1962.

M. L. Righini Bonelli and W. R. Shea, eds., *Reason, Experiment, and Mysticism: In the Scientific Revolution*. New York: Science History Publications, 1975.

E. A. Burtt, *The Metaphysical Foundations of Modern Physical Science: A Historical and Critical Essay*. New York: Harcourt, Brace & Co., 1932.

J. Campbell, ed., *Spirit and Nature: Papers from the Eranos Yearbooks*. New York: Pantheon Books (Bollingen Series XXX.1), 1954.

G. E. Christianson, *In the Presence of the Creator: Isaac Newton and His Times.* New York: Macmillan (Free Press), 1984.

A. C. Crombie, *Augustine to Gallileo:* Vol. 1, *Science in the Middle Ages, 5th to 13th Centuries*; Vol. 2, *Science in the Later Middle Ages and Early Modern Times, 13th to 17th Centuries.* London: Heinemann Educational Books, 1979.

A. G. Debus, *The Chemical Dream of the Renaissance.* Cambridge, Eng.: W. Heffer and Sons, Ltd., 1968.

———, *Man and Nature in the Renaissance.* Cambridge: Cambridge University Press, 1978 (reprinted 1981).

B. J. Teeter Dobbs, *The Foundations of Newton's Alchemy: or "The Hunting of the Greene Lyon."* Cambridge: Cambridge University Press, 1975.

C. Gillispie, ed., *Dictionary of Scientific Biography,* 15 vols. New York: Charles Scribner's Sons, 1970–1980.

A. R. Hall and M. B. Hall, *A Brief History of Science.* New York: New American Library (Signet Science Library), 1964.

B. Hansen, "The complementarity of science and magic before the scientific revolution," *American Scientist* 74 (March–April 1986): 128–36.

C. Hill, *The World Turned Upside Down: Radical Ideas During the English Revolution.* London: Temple Smith, 1972.

S. L. Jaki, *Science and Creation: From Eternal Cycles to an Oscillating Universe.* New York: Science History Publications, 1974.

W. James, *Psychology.* New York: Henry Holt & Co., 1915.

W. T. Jones, *A History of Western Philosophy.* New York: Harcourt, Brace & Co., 1952.

C. G. Jung, *The Collected Works of C. G. Jung,* trans. R. F. C. Hull. Princeton, N.J.: Princeton University Press, 1959–1983.

———, *Memories, Dreams, Reflections,* A. Jaffe, ed., trans. R. and C. Winston. New York: Random House (Vintage Books), 1965.

H. Kearney, *Science and Change 1500–1700.* New York: McGraw-Hill (World University Library), 1971.

J. E. McGuire and P. M. Rattansi, "Newton and the Pipes of Pan," *Notes and Records of the Royal Society of London* 21 (1966): 108–43.

S. F. Mason, *A History of the Sciences: Main Currents of Scientific Thought.* New York: Collier, 1962.

S. H. Nasr, *Science and Civilization in Islam.* Cambridge, Mass.: Harvard University Press, 1968.

J. Needham, *The Grand Titration: Science and Society in East and West.* Toronto: University of Toronto Press, 1969.

R. Olson, *Science Deified & Science Defied: The Historical Significance of Science in Western Culture—from the Bronze Age to the Beginnings of the Modern Era ca. 3500* B.C. *to ca.* A.D. *1640.* Berkeley and Los Angeles: University of California Press, 1982.

W. Pagel, *Paracelsus: An Introduction to Philosophical Medicine in the Era of the Renaissance,* 2nd rev. ed. Basel and New York: Karger, 1982.

I. Regardie, *The Philosopher's Stone: A Modern Comparative Approach to Alchemy from the Psychological and Magical Points of View.* St. Paul, Minn.: Llewellyn Publications, 1978.

S. E. Stumpf, *Philosophy: History and Problems.* New York: McGraw-Hill, 1971.

E. Swedenborg, *Divine Love and Wisdom.* New York: Swedenborg Foundation, 1979.

M. Teich and R. Young, eds., *Changing Perspectives in the History of Science.* London: Heinemann Educational Books, 1973.

L. Thorndike, *A History of Magic and Experimental Science,* 8 vols. New York: Columbia University Press, 1923–1958.

R. S. Westman and J. E. McGuire, *Hermeticism and the Scientific Revolution.* Los Angeles: University of California (William Andrews Clark Memorial Library), 1977.

4. PARAPSYCHOLOGICAL PERSPECTIVE

W. F. Barrett, *On the Threshold of the Unseen.* New York: E. P. Dutton, 1918.

C. D. Broad, *Religion, Philosophy and Psychical Research.* London: Routledge & Kegan Paul, 1953.

H. I. F. M. Brugmans, "A report on telepathic experiments done in the psychology laboratory at Groningen," *Le Compte Rendu Officiel du Premier Congres International des Recherches Psychiques* (Copenhagen, 1922).

H. Carrington, *The Story of Psychic Science (Psychical Research).* New York: I. Washburn, 1931.

J. E. Coover, *Experiments in Psychical Research.* Palo Alto, Calif.: Stanford University Press, 1917.

W. Crookes, *Researches into the Phenomena of Modern Spiritualism.* Rochester, N.Y.: Austin Publishing Co., 1904.

———, *Crookes and the Spirit World,* collected by R. G. Medhurst. New York: Taplinger, 1972.

G. H. Estabrooks, "A contribution to experimental telepathy," *Bulletin of the Boston Society for Psychic Research* 5 (1927): 1–30. (Reprinted in: *Journal of Parapsychology* 25 [1961]: 190–213.)

A. Gauld, *The Founders of Psychical Research.* New York: Schocken Books, 1968.

E. Gurney, F. W. H. Meyers, and F. Podmore, *Phantasms of the Living,* 2 vols. London: Trübner, 1886.

J. A. Hill, *Spiritualism: Its History, Phenomena, and Doctrine.* New York: George H. Dorian, 1919.

R. G. Jahn, "The persistent paradox of psychic phenomena: An engineering perspective," *Proceedings IEEE* 70 (February 1982): 136–170.

A. Koestler, *The Roots of Coincidence: An Excursion into Parapsychology.* New York: Random House (Vintage Books), 1972.

S. Krippner, ed., *Advances in Parapsychological Research,* 4 vols. Vols. 1–3, New York: Plenum Press, 1977–83; Vol. 4, Jefferson, N.C.: McFarland, 1984.

O. Lodge, *Phantom Walls.* New York: G. P. Putnam's Sons (Knickerbocker Press), 1930.

J. McClenon, *Deviant Science: The Case of Parapsychology.* Philadelphia: University of Pennsylvania Press, 1984.

R. A. McConnell, *Encounters with Psychology.* University of Pittsburgh: R. A. McConnell, 1981.

———, *An Introduction to Parapsychology in the Context of Science.* University of Pittsburgh: R. A. McConnell, 1983.

E. D. Mitchell, *Psychic Exploration,* J. White, ed. New York: G. P. Putnam's Sons, 1974.

G. Murphy and R. O. Ballou, *William James on Psychical Research.* New York: Viking Press, 1969.

E. K. Naumov and L. V. Vilenskaya, *Bibliographies on Parapsychology (Psychoenergetics) and Related Subjects—U.S.S.R.,* JPRS 55 (March 1972): 557, U. S. Joint Publications Research Service.

S. Ostrander and L. Schroeder, *Psychic Discoveries Behind the Iron Curtain.* New York: Prentice-Hall, 1970.

J. G. Pratt, J. B. Rhine, C. E. Stuart, and B. M. Smith, *Extrasensory Perception After Sixty Years.* New York: Holt, 1940.

K. R. Rao, ed., *The Basic Experiments in Parapyschology.* Jefferson, N.C.: McFarland, 1984.

J. B. Rhine, *Extrasensory Perception.* Boston: Boston Society for Psychic Research, 1934. Reprint. Boston: Branden Press, 1964.

J. B. Rhine and J. G. Pratt, *Parapsychology: Frontier Science of the Mind,* rev. ed. Springfield, Ill.: Charles C. Thomas, 1962.

L. E. Rhine, *Mind Over Matter: Psychokinesis.* New York: Macmillan, 1970.

W. G. Roll, *The Poltergeist.* Metuchen, N.J.: Scarecrow Press, 1976.

G. R. Schmeidler and R. A. McConnell, *ESP and Personality Patterns.* Westport, Conn.: Greenwood Press, 1973.

I. Stevenson, *Twenty Cases Suggestive of Reincarnation,* 2nd rev. ed. Charlottesville: University Press of Virginia, 1974.

L. T. Troland, *A Technique for the Experimental Study of Telepathy and Other Alleged Clairvoyant Processes,* ca. 1917. (Reprinted in: *Journal of Parapsychology* 40 [1976]: 194–216.)

F. M. Turner, *Between Science and Religion.* New Haven, Conn.: Yale University Press, 1974.

M. Ullman, S. Krippner, and A. Vaughan, *Dream Telepathy.* New York: Macmillan, 1973.

L. L. Vasiliev, *Experiments in Distant Influence: Discoveries by Russia's Foremost Parapsychologist,* A. Gregory, ed. New York: E. P. Dutton, 1983.

R. A. White, ed., *Parapsychology Abstracts International: Cumulative Author/Title/Subject Index,* vols. 1–2, 1983–1984. Dix Hills, N.Y.: Parapsychology Sources of Information Center, 1986.

R. A. White and L. A. Dale, *Parapsychology: Sources of Information.* Metuchen, N.J.: Scarecrow Press, 1973.

B. B. Wolman, ed., *Handbook of Parapsychology.* New York: Van Nostrand Reinhold, 1977.

English Language Journals: *European Journal of Parapsychology, Journal of the American Society for Psychical Research, Journal of Parapsychology, Journal of Scientific Exploration, Journal of the Society for Psychical Research, Research in Parapsychology* (Abstracts of the Proceedings of the Annual Parapsychological Association Conventions).

5. CRITICAL COUNTERPOINT

G. O. Abell and B. Singer, eds., *Science and the Paranormal.* New York: Charles Scribner's Sons, 1982.

C. Akers, "Methodological criticisms of parapsychology," in *Advances in Parapsychological Research*, vol. 4, S. Krippner, ed. Jefferson, N.C.: McFarland, 1984.

H. M. Collins and T. J. Pinch, *Frames of Meaning: The Social Construction of Extraordinary Science*. London: Routledge & Kegan Paul, 1982.

M. Gardner, *Science: Good, Bad, and Bogus*. Buffalo, N.Y.: Prometheus Books, 1981.

C. E. M. Hansel, *ESP: A Scientific Evaluation*. New York: Charles Scribner's Sons, 1966.

R. Hyman, "Parapsychological Research: A Tutorial Review and Critical Appraisal," *Proceedings IEEE* 74 (June 1986): 823–849.

R. G. Jahn, "Comments on 'Parapsychological Research: A Tutorial Review and Critical Appraisal,'" *Proceedings IEEE* 75 (April 1987): 524–525.

P. Kurtz, ed., *A Skeptic's Handbook of Parapsychology*. Buffalo, N.Y.: Prometheus Books, 1985.

D. Marks and R. Kammann, *The Psychology of the Psychic*. Buffalo, N.Y.: Prometheus Books, 1980.

F. Podmore, *Modern Spiritualism: A History and a Criticism*, vols. 1 and 2. London: Methuen, 1902.

W. F. Prince, *The Enchanted Boundary: Being a Survey of Negative Reactions to Claims of Psychic Phenomena 1820–1930*. Boston: Boston Society for Psychic Research, 1930.

Skeptical Inquirer (Journal of the Committee for the Scientific Investigation of Claims of the Paranormal). Buffalo, N.Y.

L. Zusne and W. H. Jones, *Anomalistic Psychology: A Study of Extraordinary Phenomena of Behavior and Experience*. Hillsdale, N.J.: L. Erlbaum, 1982.

6. QUANTUM CLUES

D. Bohm, *Quantum Theory*. New York: Prentice-Hall, 1951.

N. Bohr, "Light and life," *Nature* 131 (April 1933), p. 458.

———, *Atomic Physics and Human Knowledge*. New York: Wiley, 1958.

L. de Broglie, "The concepts of contemporary physics and Bergson's ideas on time and motion," in *Bergson and the Evolution of Physics*, P. A. Y. Gunter, ed. Knoxville: University of Tennessee Press, 1969.

———, *Physics and Microphysics*, trans. M. Davidson. New York: Pantheon Books, 1955.

M. Cazenave, ed., *Cosmos and Consciousness: Two Views of the Universe*. New York: Pergamon Press, 1984.

C. T. K. Chari, "Quantum mechanical perplexities," *The Philosophical Quarterly* (Bombay) 26 (October 1953): 177–185.

R. G. Colodny, ed., *Paradigm and Paradoxes, The Philosophical Challenge of the Quantum Domain.* Pittsburgh: University of Pittsburgh Press, 1978.

P. A. Degen, *Quantum Physics, Mystical Philosophy and Parapsychology: An Annotated Bibliography with an Introductory Essay.* W. Cornwall, Conn.: Locust Hill Press, in preparation.

A. Eddington, *The Philosophy of Physical Science.* New York: Macmillan, 1939.

B. d'Espagnat, *Conceptual Foundations of Quantum Mechanics*, 2nd ed. Reading, Mass.: W. A. Benjamin, 1976.

W. Heisenberg, *Physics and Philosophy: The Revolution in Modern Science.* New York: Harper & Row (Harper Torchbooks), 1958.

B. Hoffmann, *The Strange Story of the Quantum.* Magnolia, Mass.: P. Smith, 1963.

M. Jammer, *The Philosophy of Quantum Mechanics: The Interpretation of Quantum Mechanics in Historical Perspective.* New York: Wiley, 1974.

J. Mehra, ed., *The Physicist's Conception of Nature.* Boston: D. Reidel, 1973.

E. P. Wigner, "The place of consciousness in modern physics," in *Consciousness and Reality*, C. A. Muses and A. M. Young, eds. New York: Outerbridge & Lazard, 1972.

K. Wilber, ed., *Quantum Questions: Mystical Writings of the World's Great Physicists.* Boulder, Colo.: Shambhala, 1984.

7. MODERN MAN/MODERN MACHINE

A. R. Anderson, ed., *Minds and Machines.* Englewood Cliffs, N.J.: Prentice-Hall, 1964.

M. Boden, *Artificial Intelligence and Natural Man.* New York: Basic Books, 1977.

S. K. Card, T. P. Moran, and A. Newell, *The Psychology of Human–Computer Interaction.* Hillsdale, N.J.: L. Erlbaum, 1983.

E. Charniak and D. McDermott, *Introduction to Artificial Intelligence.* Reading, Mass.: Addison-Wesley, 1985.

L. S. Mark, J. S. Warm, and R. C. Huston, *Ergonomics and Human Factors.* New York: Springer-Verlag, 1987.

P. C. Jackson, Jr., *Introduction to Artificial Intelligence*, 2nd ed. New York: Dover, 1985.

A. S. Tanenbaum, *Structured Computer Organization.* Englewood Cliffs, N.J.: Prentice-Hall, 1976.

D. A. Waterman, *A Guide to Expert Systems.* Reading, Mass.: Addison-Wesley, 1986.

N. Wiener, *The Human Use of Human Beings: Cybernetics and Society.* New York: Avon Books, 1967.

8. STATISTICAL SCIENCE

G. E. P. Box, W. G. Hunter, and J. S. Hunter, *Statistics for Experimenters: An Introduction to Design, Data Analysis, and Model Building.* New York: Wiley, 1978.

W. Feller, *An Introduction to Probability Theory and Its Applications,* vol. 1, 2nd ed. New York: Wiley, 1957.

J. A. Greenwood and C. E. Stuart, "Mathematical techniques used in ESP research," *Journal of Parapsychology* 1 (1937): 206–225.

J. Marks, "The origins and development of statistics and computing," in *Science and the Making of the Modern World.* London: Heinemann, 1983, pp. 324–337.

G. W. Snedecor and W. G. Cochran, *Statistical Methods,* 7th ed. Ames, Iowa: Iowa State University Press, 1980.

S. M. Stigler, *The History of Statistics: The Measure of Uncertainty Before 1900.* Cambridge, Mass.: Harvard University Press (Belknap Press), 1986.

S. S. Wilks, "Statistical aspects of experiments in telepathy," a lecture by Samuel S. Wilks given at the Galois Institute of Mathematics, Long Island University. (Mimeographed, copyright 1938 by H. C. L. R. Lieber.)

SECTION TWO

Man/Machine Margins

R. E. Berger, "Psi Effects Without Real-Time Feedback." Science Unlimited Research Foundation, 311D Spencer Lane, San Antonio, TX 78201, 1987.

B. J. Dunne, R. G. Jahn, and R. D. Nelson, "An REG Experiment with Large Data-Base Capability." Technical Note PEAR 81001, Princeton Engineering Anomalies Research, Princeton University, School of Engineering/Applied Science, August 1981.

———, "An REG Experiment with Large Data-Base Capability, II: Effects of Sample Size and Various Operators." Technical Note PEAR 82001, Princeton Engineering Anomalies Research, Princeton University, School of Engineering/Applied Science, April 1982.

B. J. Dunne, R. D. Nelson, and R. G. Jahn, "Operator-related anomalies in a random mechanical cascade," *Journal of Scientific Exploration* 2 (1988): 155–179.

R. G. Jahn, "Psychic process, energy transfer, and things that go bump in the night," *Princeton Alumni Weekly* (4 December 1978), S-1-12.

R. G. Jahn, B. J. Dunne, and R. D. Nelson, "Engineering anomalies research," *Journal of Scientific Exploration* 1 (1987): 21–50.

R. G. Jahn, R. D. Nelson, and B. J. Dunne, "Variance Effects in REG Series Score Distributions." Technical Note PEAR 85001, Princeton Engineering Anomalies Research, Princeton University, School of Engineering/Applied Science, July 1985.

E. C. May, B. S. Humphrey, and G. S. Hubbard, "Electronic System Perturbation Techniques," SRI International, Menlo Park, Calif., Technical Report, September 30, 1980.

R. D. Nelson, B. J. Dunne, and R. G. Jahn, "A Psychokinesis Experiment with a Random Mechanical Cascade." Technical Note PEAR 83002, Princeton Engineering Anomalies Research, Princeton University, School of Engineering/Applied Science, June 1983.

D. I. Radin and R. D. Nelson, "Evidence for consciousness-related anomalies in random physical systems," *Foundations of Physics* 19 (December 1989): in press.

SECTION THREE

Precognitive Remote Perception

S. Allen, P. Green, K. Rucker, R. Cohen, C. Goolsby, and R. L. Morris, "A remote viewing study using a modified version of the SRI procedure," in *Research in Parapsychology 1975*. Metuchen, N.J.: Scarecrow Press, 1976, pp. 46–48.

H. Chotas, "Remote viewing in the Durham area," *Journal of Parapsychology* 1 (1978): 61–62.

B. J. Dunne, *Precognitive Remote Perception.* M.A. thesis, Committee on Human Development, University of Chicago, 1979.

———, "Long distance precognitive remote viewing," in *Research in Parapsychology 1978.* Metuchen, N.J.: Scarecrow Press, 1979, pp. 68–70.

———, "Multiple subject precognitive remote viewing," in *Research in Parapsychology 1977.* Metuchen, N.J.: Scarecrow Press, 1978, pp. 146–151.

B. J. Dunne and J. P. Bisaha, "Precognitive remote perception: A critical overview of the experimental program," in *Research in Parapsychology 1979.* Metuchen, N.J.: Scarecrow Press, 1980, pp. 117–120.

———, "Precognitive remote viewing in the Chicago area," *Journal of Parapsychology* 43 (1979): 17–30.

J. W. Dunne, *An Experiment with Time.* London: Farber and Farber, 1939.

A. C. Hastings and D. B. Hurt, "A confirmatory remote viewing experiment in a group setting," *Proceedings IEEE* 64 (1976): 1544–1545.

R. G. Jahn, B. J. Dunne, R. D. Nelson, E. G. Jahn, T. A. Curtis, and I. A. Cook, "Analytical judging procedure for remote perception experiments—II: Ternary coding and generalized descriptors," Technical Note PEAR 82002, Princeton Engineering Anomalies Research, Princeton University, School of Engineering/Applied Science, 1982.

R. G. Jahn, B. J. Dunne, and E. G. Jahn, "Analytical judging procedure for remote perception experiments," *Journal of Parapsychology* 3 (1980): 207–231.

E. Karnes, J. Ballou, E. Sussman, and F. Swaroff, "Remote viewing: Failures to replicate with control comparisons," *Psychological Reports* 45 (1979): 963–973.

D. F. Marks, "Investigating the Paranormal," *Nature* 320 (1986): 119–124.

D. F. Marks and R. Kammann, Letter, *Nature* 274 (1978): 680–681.

R. L. Morris, "An exact method for evaluating preferentially matched free-response material," *Journal of the American Society for Psychical Research* 66 (1972): 401.

H. E. Puthoff and R. Targ, *Mind-Reach.* New York: Delacorte Press, 1977.

H. E. Puthoff, R. Targ, and E. C. May, "Experimental psi research:

Implication for physics," in *The Role of Consciousness in the Physical World,* R. G. Jahn, ed. Boulder, Colo.: Westview Press, 1981, pp. 37–86.

M. Schlitz and E. Gruber, "Transcontinental remote viewing," *Journal of Parapsychology* 4 (1980): 305–317.

Upton Sinclair, *Mental Radio.* New York: Collier, 1971.

C. Scott, "On the evaluation of verbal material in parapsychology: A discussion of Dr. Pratt's monograph," *Journal of the Society for Psychical Research* 46 (1972): 79–90.

G. Solfvin, E. Kelly, and D. Burdick, "Some new methods of analysis for preferential ranking data," *Journal of the American Society for Psychical Research* 72 (1978): 93.

R. Targ and H. E. Puthoff, "Information transmission under conditions of sensory shielding," *Nature* 252 (1974): 602–607.

R. Targ and K. Harary, *The Mind Race: Understanding and Using Psychic Abilities.* New York: Villard Books, 1984.

C. T. Tart, H. E. Puthoff, and R. Targ, eds., *Mind at Large: IEEE Symposia on the Nature of Extrasensory Perception.* New York: Praeger Special Studies, 1979.

SECTION FOUR

The Waves of Consciousness

1. MODEL MAZE

D. Bohm, *Wholeness and the Implicate Order.* New York: Routledge & Kegan Paul, 1980.

P. Diaconis, "Statistical problems in ESP research," *Science* 201 (14 July 1978): 131–136.

L. L. Gatlin, *Information Theory and the Living System.* New York: Columbia University Press, 1972.

R. G. Jahn, ed., *The Role of Consciousness in the Physical World.* Boulder, Colo.: Westview Press, 1981.

I. M. Kogan, "Information theory analysis of telepathic communications experiments," *Radio Engineer* 23 (1968): 122.

K. Kornwachs and W. von Lucadou, "Development of the system theoretic approach to psychokinesis," *European Journal of Parapsychology* 3 (1980): 297–314.

L. Oteri, ed., *Quantum Physics and Parapsychology.* New York: Parapsychology Foundation, 1975.

A. Puharich, ed., *The Iceland Papers: Select Papers on Experimental and Theoretical Research on the Physics of Consciousness.* Amherst, Wisc.: Essentia Research Associates, 1979.

M. Ryzl, "A model for parapsychological communication," *Journal of Parapsychology* 30 (1966): 18–31.

B. Shapin and L. Coly, eds., *Concepts and Theories of Parapsychology.* New York: Parapsychology Foundation, 1981.

R. G. Stanford, "An experimentally testable model for spontaneous psi events; I. Extrasensory events," *Journal of the American Society for Psychical Research* 68 (1974): 34–57. "II. Psychokinetic events," *Ibid.*, pp. 321–356.

H. P. Stapp, "Mind, matter and quantum mechanics," *Foundations of Physics* 12 (April 1982): 363–399.

A. M. Young, *The Geometry of Meaning.* New York: Delacorte Press (A Merloyd Lawrence Book), 1976.

2. REALITY RECIPE

R. G. Jahn and B. J. Dunne, "On the Quantum Mechanics of Consciousness with Application to Anomalous Phenomena." Technical Note PEAR 83005.1, Princeton Engineering Anomalies Research, Princeton University, School of Engineering/Applied Science, 1984.

———, "On the quantum mechanics of consciousness, with application to anomalous phenomena," *Foundations of Physics* 16 (August 1986): 721–772.

C. A. Muses and A. M. Young, *Consciousness and Reality.* New York: Outerbridge & Lazard, 1972.

3. CONSCIOUSNESS PARTICLES AND WAVES

W. C. Elmore, *Physics of Waves.* New York: McGraw-Hill, 1969.

R. P. Feynman, R. B. Leighton, and M. Sands, *The Feynman Lectures on Physics,* vol. 3. Reading, Mass.: Addison-Wesley, 1965.

S. Leibovich and A. R. Seebass, eds., *Non-Linear Waves.* Ithaca: Cornell University Press, 1974.

M. J. Lighthill, *Waves in Fluids.* New York: Cambridge University Press, 1978.

H. Pain, *The Physics of Vibrations and Waves.* London: Wiley, 1968.

C. H. Papas, *Theory of Electromagnetic Wave Propagation.* New York: McGraw-Hill, 1965.

J. P. Pierce, *Almost All About Waves.* Cambridge, Mass.: MIT Press, 1974.

F. K. Richtmyer, E. H. Kennard, and J. N. Cooper, *Introduction to Modern Physics,* 6th ed. New York: McGraw-Hill, 1969.

J. A. Stratton, *Electromagnetic Theory.* New York: McGraw-Hill, 1941.

R. A. Waldron, *Waves and Oscillations.* Princeton, N.J.: D. van Nostrand, 1964.

G. B. Whitham, *Linear and Non-linear Waves.* New York: Wiley, 1974.

4. QUANTUM CONCEPTS

N. Bohr, "Causality and complementarity," *Philosophy of Science* 4 (July 1937): 289–298.

M. Born, *Atomic Physics.* New York: Hafner, 1946.

H. Goldstein, *Classical Mechanics.* Cambridge, Mass.: Addison-Wesley, 1950.

K. Gottfried, *Quantum Mechanics.* New York: W. A. Benjamin, 1966.

W. Heisenberg, *The Physical Principles of the Quantum Theory,* trans. C. Eckart and F. C. Hoyt. New York: Dover, 1949.

L. I. Schiff, *Quantum Mechanics.* New York: McGraw-Hill, 1949.

J. von Neumann, *Mathematical Foundations of Quantum Mechanics,* trans. R. T. Beyer. Princeton, N.J.: Princeton University Press, 1955.

V. Weisskopf, *Knowledge and Wonder: The Natural World as Man Knows It.* Cambridge, Mass.: MIT Press, 1979.

5. CONSCIOUSNESS COORDINATES

S. H. Booth, ed., *Precision Measurement and Calibration,* 3 vols. Washington: National Bureau of Standards, 1961.

A. A. Burr, *Measurements in Applied Physics.* London: Chapman & Hall, 1968.

J. Campbell, ed., *Man and Time: Papers from the Eranos Yearbooks.* New York: Pantheon Books (Bollingen Series XXX.3), 1957.

J. B. Carroll, ed., *Language, Thought, and Reality: Selected Writings of Benjamin Lee Whorf.* Cambridge: MIT Press, 1987.

P. C. W. Davies, *The Physics of Time Asymmetry.* Berkeley and Los Angeles: University of California Press, 1977.

S. J. Dimond and J. G. Beaumont, *Hemisphere Function in the Human Brain.* New York: Wiley (Halsted Press), 1974.

M. S. Gazzaniga, "Right hemisphere language following brain bisection: A 20-year perspective," *American Psychologist* 38 (May 1983): 525–549.

G. P. Harnwell, *Experimental Atomic Physics.* New York: McGraw-Hill, 1933.

E. Husserl, *The Phenomenology of Internal Time Consciousness.* Bloomington: Indiana University Press, 1984.

J. E. Le Doux and W. Hirst, eds., *Mind and Brain: Dialogues in Cognitive Neuroscience.* New York: Cambridge University Press, 1986.

G. A. Miller and P. N. Johnson-Laird, *Language and Perception.* Cambridge, Mass.: Harvard University Press (Belknap Press), 1976.

A. Ortony, ed., *Metaphor and Thought.* Cambridge: Cambridge University Press, 1979.

G. J. Whitrow, *The Natural Philosophy of Time.* Oxford: Clarendon Press, 1980.

R. E. Zupko, *A Dictionary of English Weights and Measures; from Anglo-Saxon Times to the Nineteenth Century.* Madison: University of Wisconsin Press, 1968.

6. CONSCIOUSNESS ATOMS

E. H. Erikson, *Childhood and Society.* New York: Norton, 1963.

S. Freud, *New Introductory Lectures on Psychoanalysis.* New York: Norton, 1933.

G. Herzberg, *Atomic Spectra and Atomic Structure.* New York: Dover, 1945.

C. G. Jung, *Psychological Types.* New York: Harcourt Brace and Co., 1923.

L. Kohlberg, "Development of moral character and moral ideology," in *Review of Child Development Research,* vol. 1, M. L. Hoffman and L. W. Hoffman, eds. New York: Russell Sage Foundation, 1964.

C. W. Leadbetter, *The Chakras.* Wheaton, Ill.: Theosophical Publishing House, 1972.

A. H. Maslow, *Motivation and Personality.* New York: Harper & Row, 1954.

G. Murchie, *Music of the Spheres: The Material Universe from Atom to Quasar, Simply Explained,* vol. 2. New York: Dover, 1961.

J. Piaget, *The Origins of Intelligence in Children.* New York: International Universities Press, 1952.

7. CONSCIOUSNESS MOLECULES

D. R. Bates, *Atomic and Molecular Processes.* New York: Academic Press, 1962.

M. E. Dyatkina and Y. K. Syrkin, *The Structure of Molecules and the Chemical Bond.* New York: Interscience Publishers, 1950.

W. Heitler, *Elementary Wave Mechanics.* Oxford: Clarendon Press, 1948.

E. B. Taub-Bynum, *The Family Unconscious: "An Invisible Bond."* Wheaton, Ill.: The Theosophical Publishing House (Quest), 1984.

J. H. Woods, trans., *The Yoga System of Pantanjali.* Cambridge, Mass.: Harvard Oriental Series XVII, 1924.

8. METAPHORICAL PRINCIPLES

F. Capra, *The Tao of Physics.* Boulder, Colo.: Shambhala, 1974.

P. Davies, *Other Worlds: Space, Superspace and the Quantum Universe.* New York: Simon & Schuster (Touchstone), 1980.

R. Jones, *Physics as a Metaphor.* New York: University of Michigan Press, 1978.

G. Zukav, *The Dancing Wu Li Masters: An Overview of the New Physics.* New York: W. Morrow, 1979.

11. BROADER BASIS

C. S. Cook, *Structure of Atomic Nuclei.* Princeton, N.J.: D. van Nostrand, 1964.

R. Fowler and E. A. Guggenheim, *Statistical Thermodynamics.* Cambridge: Cambridge University Press, 1952.

H. Fruuenfelden and E. Henley, *Subatomic Physics.* New York: Prentice-Hall, 1974.

D. Halliday, *Introductory Nuclear Physics.* New York: Wiley, 1950.

C. Hampden-Turner, *Maps of the Mind.* New York: Macmillan, 1982.

C. G. Jung, *The Archetypes and the Collective Unconscious,* trans. R. F. C. Hull. Princeton, N.J.: Princeton University Press, 1969.

D. H. Perkins, *Introduction to High Energy Physics,* 2nd ed., rev., enl., and reset. Reading, Mass.: Addison-Wesley Advanced Book Program (World Science Division), 1982.

E. Segré, *Nuclei and Particles: An Introduction to Nuclear and Sub-nuclear Physics.* New York: W. A. Benjamin, 1965.

SECTION FIVE

The Vectors Deflected

2. MAN/MACHINE MARGINS (See also Section One, Chapter 7)

S. C. Florman, *The Existential Pleasures of Engineering.* New York: St. Martin's Press, 1976.

I. Peterson, "Warning: This software may be unsafe," *Science News* 130 (13 September 1986): 171–173.

3. PARAPSYCHOLOGICAL PROGRESS (See also Section One, Chapter 4)

D. Dean, J. Mihalasky, S. Ostrander, and L. Schroeder, *Executive ESP.* Englewood Cliffs, N.J.: Prentice-Hall, 1974.

R. G. Jahn, "On the Representation of Psychic Research to the Community of Established Science," invited address in *Research in Parapsychology 1983: Abstracts and Papers from the Twenty-sixth Annual Convention of the Parapsychological Association, 1983*, R. A. White and R. S. Broughton, eds. Metuchen, N.J.: Scarecrow Press, 1984.

M. Johnson (chairperson), "Symposium: Current Directions in European Parapsychology" in *Research in Parapsychology 1978: Abstracts and Papers from the Twenty-first Annual Convention of the Parapsychological Association, 1978*, W. G. Roll, ed. Metuchen, N.J.: Scarecrow Press, 1979.

J. Ludwig, ed., *Philosophy and Parapsychology.* Buffalo, N.Y.: Prometheus Books, 1978.

E. C. May, D. I. Radin, G. S. Hubbard, B. S. Humphrey, and J. M. Utts, "Psi experiments with random number generators: An informational model." Proceedings of the Presented Papers of the 28th Annual Parapsychological Assn. Convention, Tufts University, Medford, Mass., August 12–16, 1985.

I. M. Owen, with M. Sparrow, *Conjuring Up Philip: An Adventure in Psychokinesis.* New York: Pocket Books (A Kangaroo Book), 1977.

"Research on the Physics of Consciousness (Parapsychology)" in *Survey of Science and Technology Issues Present and Future* (Staff Report of the Committee on Science and Technology, U.S. House of Representatives, Ninety-seventh Congress, first session, June 1981).

R. Rosenthal, *Meta-Analytic Proceedings for Social Research.* Applied Social Research Methods Series, vol. 6. Beverly Hills, Calif.: Sage Publications, 1984.

G. R. Schmeidler, ed., *Parapsychology: Its Relation to Physics, Biology, Psychology and Psychiatry.* Metuchen, N.J.: Scarecrow Press, 1976.

Z. Vassy, "Theoretical and methodological considerations on experiments with pseudorandom number sequences," *Journal of Parapsychology* 49 (1985): 127–153.

4. CRITICAL COLLOQUY (See also Section One, Chapter 5)

"The Ganzfeld Debate," *Journal of Parapsychology* 50 (December 1986, entire issue).

J. Alcock, "Parapsychology: Science of the anomalous or search for the soul?" *The Behavioral and Brain Sciences,* March 1988, in press.

K. R. Rao and J. Palmer, "The Anomaly of Psi: Recent research and criticism." *The Behavioral and Brain Sciences,* March 1988, in press.

5. STATISTICAL SENSITIVITIES (See also Section One, Chapter 8, and Section Four, Chapter 11)

"The Concept of Probability," Proceedings of the Third International Meeting on Epistemology, Delphi, Greece, October 1987; organized by the Group of Interdisciplinary Research (Dept. of Physics, University of Athens).

G. S. Brown, *Probability and Scientific Inference.* New York: Longmans, Green, 1957.

N. Davidson, *Statistical Mechanics.* New York: McGraw-Hill, 1962.

R. C. Tolman, *The Principles of Statistical Mechanics.* New York: Dover, 1979.

6. QUANTUM CONSEQUENCES (See also Section One, Chapter 7, and Section Four)

J. D. Barrow and F. J. Tipler, *The Anthropic Cosmological Principle.* New York: Oxford University Press, 1986.

J. S. Bell, "On the Einstein-Podolsky-Rosen paradox," *Physics* 1 (1964): 195–200.

D. Bohm and J. Bub, "A proposed solution of the measurement problem in quantum mechanics by a hidden variable theory," *Review of Modern Physics* 38 (1966): 453.

J. F. Clauser and A. Shimony, "Bell's theorem: Experimental tests and implications," *Reports on Progress in Physics* 41 (1978): 1881.

O. Costa de Beauregard, "Time symmetry and interpretation of quantum mechanics," *Foundations of Physics* 6 (1976): 539–559.

B. d'Espagnat, *In Search of Reality.* Berlin: Springer-Verlag, 1983.

S. W. Hawking, "The quantum mechanics of black holes," *Scientific American* (January 1977): 34–40.

N. Herbert, *Quantum Reality: Beyond the New Physics.* New York: Doubleday (Anchor Press), 1985.

M. B. Hesse, *Forces and Fields: The Concept of Action at a Distance in the History of Physics.* New York: Philosophical Library, 1962.

H. Margenau, "ESP in the framework of modern science," *Journal of the American Society for Psychical Research* 60 (1966): 214–228.

J. Mehra, "Quantum mechanics and the explanation of life," *American Scientist* 61 (November–December 1973): 722–728.

N. D. Mermin, "Is the moon there when nobody looks? Reality and the quantum theory," *Physics Today* (April 1985): 38–47.

J. V. Narlikar, "Cosmic Tachyons: An astrophysical approach," *American Scientist* 66 (1978): 588–593.

Y. F. Orlov, "The wave logic of consciousness: A hypothesis," *International Journal of Theoretical Physics* 21 (1982): 37–53.

F. Rohrlich, "Facing quantum mechanical reality," *Science* 221 (1983): 1251.

L. L. Smarr and W. H. Press, "Our elastic spacetime: Black holes and gravitational waves," *American Scientist* 66 (January–February 1978): 72–79.

D. E. Thomsen, "Anomalons get more and more anomalous," *Science News* 125 (25 February 1984): 118.

F. A. Wolf, *Star Wave* $\psi^*\psi$*: Mind, Consciousness, and Quantum Physics.* New York: Macmillan, 1984.

7. SCHOLARLY STRETCH

P. Berger and T. Luckmann, *The Social Construction of Reality.* Garden City, N.Y.: Doubleday, 1966.

F. E. Bloom, A. Lazerson, and L. Hofstadter, *Brain, Mind and Behavior.* New York: W. H. Freeman, 1985.

N. Cousins, *The Healing Heart.* New York: Avon, 1983.

M. Delbrück, *Mind from Matter? An Essay on Evolutionary Epistemology.* Palo Alto, Calif.: Blackwell Scientific Publications, 1986.

L. Dossey, *Beyond Illness: Discovering the Experience of Health.* Boulder, Colo.: Shambhala, 1984.

J. Eccles, ed., *Mind and Brain: The Many-Faceted Problem.* New York: Paragon House, 1982.

G. Globus, ed., *Consciousness and the Brain.* New York: Plenum Press, 1976.

E. Goffman, *Frame Analysis.* New York: Harper & Row (Colophon Books), 1974.

M. Ishikawa, "Consciousness: A parameter for understanding man and nature," in *The World and I* (A publication of The Washington Times Corporation, 450 5th St. N.W., Suite 1120, Washington, DC 20001), January 1986.

R. G. Jahn, "Psychic phenomena," in *Encyclopedia of Neuroscience.* Cambridge, Mass.: Birkhäuser Boston, 1986: 993–96.

B. D. Josephson and V. S. Ramachandran, *Consciousness and the Physical World: Edited Proceedings of an Interdisciplinary Symposium on Consciousness Held at the University of Cambridge in January 1978,* 1st ed. Oxford and New York: Pergamon Press, 1980.

A. Koestler, *Janus: A Summing Up.* New York: Random House, 1978.

A. Korzybski, *Science and Sanity: An Introduction to Non-Aristotelian Systems and General Semantics.* Clinton, Mass.: Colonial Press, 1933.

E. Kubler-Ross, *On Death and Dying.* New York: Macmillan, 1969.

T. Luckmann, *The Invisible Religion.* New York: Macmillan, 1967.

H. Margenau, *The Miracle of Existence.* Woodbridge, Conn.: Ox Bow Press, 1984.

R. May, *Love and Will.* New York: Dell Publishing (A Delta Book), 1969.

W. Penfield, *The Mystery of the Mind: A Critical Study of Consciousness and the Human Brain.* Princeton, N.J.: Princeton University Press, 1975.

G. C. Quarton, T. Melnechuk, and F. O. Schmitt, eds., *The Neurosciences: A Study Program.* New York: Rockefeller University Press, 1967.

K. Ring, *Life at Death: Investigation of the Near-Death Experience.* New York: Quill, 1982.

J. Salk, *Anatomy of Reality: Merging of Intuition and Reason.* New York: Praeger Publishers (Convergence Series), 1985.

B. S. Siegel, *Love, Medicine and Miracles.* New York: Harper & Row, 1986.

R. Sperry, *Science and Moral Priority: Merging Mind, Brain and Human Values.* New York: Columbia University Press, 1983.

A. N. Whitehead, *Science and the Modern World.* Cambridge: 1926. Reprint. New York: New American Library, 1958.

L. White, B. Tursky, and G. Schwartz, eds., *Placebos: Theory, Research and Mechanisms.* New York: Guilford, 1985.

D. P. Wirth, "Healing Expectations." Masters of Science thesis, John F. Kennedy University, Orinda, CA, 1987.

8. MODERN MYSTICAL MAN (See also Section One, Chapter 2)

G. Bateson, *Mind and Nature: A Necessary Unity.* New York: Bantam Books, 1983.

A. Comfort, *Reality and Empathy: Physics, Mind and Science in the 21st Century.* Albany, N.Y.: SUNY Press, 1984.

S. Grof, ed., *Ancient Wisdom and Modern Science.* Albany, N.Y.: SUNY Press, 1984.

The Isthmus Foundation Lectures on Science and Religion, published in *Perkins Journal* 36 (Summer 1983) and 39 (April 1986), The Perkins School of Theology, Southern Methodist University, Dallas, Texas.

G. Krishna, *The Biological Basis of Religion and Genius,* with an introduction by C. F. von Weizsäcker. New York: Harper & Row, 1972.

J. Krishnamurti and D. Bohm, *The Ending of Time.* San Francisco: Harper & Row, 1985.

A. de Riencart, *The Eye of Shiva: Eastern Mysticism and Science.* New York: Morrow Quill Paperbacks, 1980.

D. L. Schindler, ed., *Beyond Mechanism: The Universe in Recent Physics and Catholic Thought.* Lanham, Md.: University Press of America, 1986.

H. Smith, *Beyond the Post-Modern Mind.* New York: Crossroads Press, 1982.

P. Teilhard de Chardin, *The Future of Man,* trans. N. Denny. New York: Harper & Row (Harper Torchbooks), 1964.

9. ORGANIZATION AND ORDER

E. Broda, *Ludwig Boltzmann: Man-Physicist-Philosopher.* Woodbridge, Conn.: Ox Bow Press, 1983.

E. Jantsch, *The Self-Organizing Universe: Scientific and Human Implications of the Emerging Paradigm of Evolution.* New York: Pergamon Press, 1983.

K. Kornwachs and W. von Lucadou, "Psychokinesis and the concept of complexity," *Psychoenergetics* 3 (1979): 327–342.

J. Monod, *Chance and Necessity.* New York: Random House (Vintage Books), 1971.

C. E. Shannon and W. Weaver, *The Mathematical Theory of Communication.* Urbana, Ill.: University of Illinois Press, 1949.

ACKNOWLEDGMENTS

The Princeton Engineering Anomalies Research (PEAR) program has been supported over the past decade by enduring major grants from the James S. McDonnell Foundation and the John E. Fetzer Foundation and by gifts from various other philanthropic organizations and individuals. This book was specifically underwritten by a generous grant from Mr. Trammell Crow. Beyond these benefactors, the authors are deeply indebted to many other friends and colleagues who have supported this work in numerous explicit and subtle ways.

Within our laboratory, Susan Intner, Roger Nelson, and many other members of the PEAR family, past and present, have been extraordinarily helpful in preparation and proofreading of the manuscript, illustrations, tables, and references, and in countless other tedious tasks. We are particularly indebted to Yolanda Pastor for presiding over the word processing of the interminable drafts with extraordinary skill, patience, and unflagging good humor. Beyond our immediate group, several members of the faculty and staff of the School of Engineering and Applied Science have selflessly provided various crucial services and personal encouragement for this book, and for our research program as a whole: Assistant Deans George Mueller and Jeremiah Farrington, Professor William Surber, and the staffs of our central maintenance, machine shop, and design groups, to name only a few.

The book could not have been completed without uncommon editorial support from its publishers, beginning with its first conceptualization and following through its various stages of design and composition. In particular, we are most grateful for the visionary strength and immense personal and professional encouragement of William Jovanovich and Peter Jovanovich, and for the technical skill, good judgment, and boundless tolerance of John Radziewicz.

The list of other friends and supporters is far too extensive to delineate in full, but specific mention must be made, for a broad range of reasons, of the very special roles of Mr. John E. Fetzer, the late James S. McDonnell, Mr. and Mrs. John F. McDonnell, Mr. George L. Ohrstrom, Jr., Mr. Charles E. Spence, Mr. Donald C. Webster, and Mr. Michael Witunski. We must also gratefully acknowledge the countless research colleagues in this and related fields, both in this country and throughout the Eastern and Western scholarly worlds, who have shared, criticized, encouraged, and just simply acknowledged our research over the past several years. Our sense of their genuine interest and selfless enthusiasm has been a monumental factor in stimulating and sustaining this effort.

And, of course, we are deeply indebted to our families for the patience with which they have endured our incessant preoccupation and protracted neglect, and have borne all the extra burdens that have been shifted to them during this long period of composition. Similar thanks extend to Rose Rackley and to others of our extended family who have buttressed our convictions on these difficult matters with their own, at a time when it was not always popular to do so. Nor should we forget the pioneers of the past, and the heirs of the future, who earlier began, and later will carry on, this expedition into the realm of creative consciousness.

Finally, the authors record their profound gratitude to the many necessarily anonymous experimental operators who have so generously given of their time, their energy, their confidence, and their love, to generate the research data on which this program, this book, and this message ultimately must stand or fall.

Robert G. Jahn and Brenda J. Dunne
Princeton, New Jersey, June 21, 1987

INDEX

b

c

d

e

f

g

h

i

j

k

l

m

P

q

r

s

V

W

y

z